LEARNING REMOTE SENSING
USING DRAGON/IPS® AND OPENDRAGON™

Kurt T. Rudahl
Sally E. Goldin

Global Software Institute October 25, 2011

Revised October 25, 2011

About the Authors:

Kurt T. Rudahl holds a Bachelors degree in mathematics from New York University and a Masters in computer science from the University of Wisconsin Madison. He spent many years as a consultant developing both custom hardware and software for a wide range of U.S. companies. Along with Dr. Goldin, he created Dragon/ips, the first remote sensing image processing system to run on off-the-shelf personal computers.

Mr. Rudahl spent two years in Thailand as an Assistant Professor at the Asian Institute of Technology, teaching in the Asian Regional Remote Sensing Training Center. He has served in the Department of Computer Engineering at King Mongkut's University of Technology Thonburi (KMUTT) in Thailand since 2003.

Sally E. Goldin earned her Bachelors and Masters degrees from Brown University and her Ph.D. from Carnegie-Mellon University. After spending several years as a researcher at the Rand Corporation, she turned her attention to real-world software development. She has more than twenty years experience as a software engineer and project leader, working with both start-ups and major U.S. corporations. Her areas of special expertise include user interface design, software process paradigms, and geoinformatics algorithms.

In the mid-nineteen-eighties, Dr. Goldin was seconded by the U.S. Agency for International Development to teach geographical applications of digital image processing theory and practice at the Asian Institute of Technology in Bangkok, Thailand. Since 2003, she has been a member of the faculty in the Department of Computer Engineering at KMUTT, teaching, advising, and conducting research.

About GSI

The Global Software Institute exists to apply computer technology to the meeting of global human needs. GSI focuses on appropriate, computer-based solutions to problems experienced by lesser-developed countries, and by disadvantaged minority communities within developed countries. Target areas include improvement of primary and secondary education, human resource development, public health, natural resource management, disaster prevention and mitigation, and the preservation of archaeological and cultural heritage. GSI serves as an expert consultant, assistant, or coach to government agencies and non-governmental organizations (NGOs) who are concerned with these and related issues.

GSI's goals are to assist and work with these organizations in realizing the potentials of computer technology, and to educate them concerning its capabilities and pitfalls. GSI seeks out and participates in cooperative projects initiated by governments and NGOs. On occasion, GSI also initiates projects to exploit new possibilities for computer-based technology in improving social welfare and economic development.

The Global Software Institute is a non-profit, 501(c)3 corporation incorporated in Massachusetts, U.S.A.

3.141

PREFACE

Geospatial knowledge has become important in almost every profession, from agriculture to zoology, and in private life as well, for planning vacations, buying a new home, or just entertainment. Satellite images have become familiar to most people due to their extensive use in the media and as background for mapping on the World Wide Web. Images from numerous satellite platforms and aerial surveys are now available - sometime free, sometimes at very high cost.

Web-based software for viewing image data is also widely available. However, much more value can be extracted from image data than simple viewing software can provide. Unfortunately, the software for more advanced processing of *remote sensing* data tends to be both expensive and to require considerable expertise.

While mission-critical remote sensing data analysis is too important to be entrusted to amateurs, there is a growing need for practitioners in very diverse occupations to be able to use geospatial data intelligently. This book, and the **Dragon/ips**® software it uses, are an effort to make that possible. This book offers a straightforward, hands-on treatment of the core concepts in remote sensing and raster GIS analysis. Readers with limited knowledge of physics or geography can use this book not only to acquire an understanding of remote sensing fundamentals but also to begin applying geospatial analysis to their own problems.

Background

Simplicity, ease of use and self-documentation have been critical objectives for **Dragon** ever since the software was created. Over the years, however, it has become clear that just being user friendly is not enough. There remains a need for a practically-oriented textbook with example exercises that illustrate how the powerful analysis capabilities of **Dragon** can be applied. For one thing, it is often the case that remote sensing techniques are taught by someone whose education predated easily-available remote sensing, and who therefore has never had personal experience in using these technologies. This book is designed to address this need, providing a single, concise, self-contained volume that combines basic theory with practice.

Since the time when Dragon was primarily an educational tool, it has evolved in several directions. The original version became the **Dragon Academic** edition. Thanks to generous support from King Mongkut's University of Technology Thonburi in Thailand, the Academic Edition became the basis for the free and open source **OpenDragon** (*www.open-dragon.org*). Meanwhile, a **Dragon Professional** edition was created to serve the needs of the advanced researcher (*www.dragon-ips.com*). Everything in this book, unless specifically noted, applies to all of these versions. The exercises use data provided at the OpenDragon web site.

There are two additional components of Dragon which are not discussed in this book: the **Programmers Toolkit**, which permits extending Dragon's functionality, and the **Dragon engine** which allows Dragon's analysis capabilities to be embedded into another system.

Audience

This is a textbook, so obviously the audience is people who want to learn about remote sensing analysis techniques. The primary targets are undergraduate students in the many disciplines that now utilize remote sensing as a standard tool. However, the book may also be appropriate for graduate students new to digital image analysis.

Many secondary school systems are now introducing remote sensing analysis as part of their geography curriculum. This book may be especially valuable for use by secondary school teachers who wish to quickly become conversant with these technologies.

As noted above, the book is intended to be self-contained (in combination with the free OpenDragon software) and is therefore entirely appropriate for individual study.

One reasonable question is whether this book would be useful to someone who has a remote sensing software package other (and possibly more sophisticated) than Dragon. Because Dragon was designed to focus on the most important and commonly used operations rather than on more exotic techniques, and because all of the analysis processes are explained in detail, the book may have a value to those using some other software system as well. However, the exercises will certainly be easier to perform by using Dragon.

Organization of the Book

Chapter 1 introduces the basic notions of remote sensing sensors, data sets, and analysis. These are general principles which will apply in any discussion of digital analysis of remotely sensed images.

Chapter 2 introduces Dragon and describes some important operating concepts that the reader must understand to use Dragon effectively. It also discusses in detail the various ways in which a user interacts with Dragon to perform image processing tasks.

Chapters 3 through 6 describe the four most substantive functional divisions in Dragon: display operations, enhancement operations, classification operations, and geography operations. Chapter 7 provides a combined treatment the utility operations and file handling operations. Each of these five chapters begins with a general overview that introduces the important image processing concepts associated with that functional division. This is followed by separate sections for each operation. These individual sections describe what the operation does and make suggestions for its most effective use. Each chapter ends with a set of exercises that give hands-on practice in using the system to achieve the goals associated with that part of the system.

The sample imagery used in the exercises can be downloaded from **http://www.open-dragon.org/teaching_mat.html**, in the file called **BookSamples.zip**. This zip file should be unpacked into the default input data directory that the user specified when installing Dragon. Some of the images are also included as part of the **OpenDragon** and **Dragon Professional** distribution, but are duplicated in the booksamples archive for convenience.

Errata and Corrections

It is said that "all software has bugs", but books can have bugs as well. Any errors we find, or that you find and send to us, will be reported on the Dragon website: **http://www.dragon-ips.com/product_documentation.html**.

Disclaimer

We wish to make it clear that this is **not** a textbook on geography, nor even on how to apply remote sensing analysis techniques to geographic problems. (We do plan a successor book, written by a professional geographer, containing laboratory exercises in applying remote sensing to real-world problems. However, that is not this book.) Rather, this book is a textbook about the "nuts and bolts" of performing analysis on remote sensing image data.

Acknowledgments

We wish to thank the various data providers who have given us permission to use their data sets, in particular the Geoinformatics and Space Development Agency (GISTDA) (Figures 1.6 and 3.2), SPOT Image Corporation (Figure 2.3) and the U.S. Geological Survey (USGS).

Contents

Chapter 1

INTRODUCTION TO REMOTE SENSING

This chapter introduces some of the fundamental concepts of remote sensing and its sister technology, geographic information systems (GIS). Both of these technologies use computers to manipulate information about the state of the earth (or occasionally, some other planet) at different locations. Together, the two technologies can help people evaluate situations, predict future changes, make decisions and develop plans. Remote sensing and GIS are two examples of *geospatial information processing*, sometimes known as *geoinformatics*. Over the past thirty years geoinformatics has become increasingly important in a wide range of human activities including agriculture, natural resource inventory and management, urban planning, disaster response, health assessment and epidemiology, and business development.

The objective of this chapter is to provide sufficient background for readers to begin hands-on experimentation with **Dragon**. For a more detailed discussion of the principles of remote sensing, readers can consult some of the texts in the *Bibliography*. These references cover the physics and mathematics behind remote sensing and GIS. However, they do not provide practical experience in working with remotely sensed data, which is the primary goal of this book.

1.1   Remote Sensing Image Analysis

Remote sensing is gathering information about some object from a distance. In most common uses of remote sensing, the object is the earth, or some part of it, and the information gathered is measured levels of electromagnetic radiation reflected or emitted from the earth at various locations. A variety of sensing instruments can be used to measure and record this radiation, depending on its wavelength. The most commonly-used sensors are aircraft-borne cameras and scanning radiometers mounted on satellites orbiting the earth.

Remotely-sensed data are routinely used for investigation, management and monitoring in a wide variety of fields: meteorology, forestry, oceanography, exploration geology, hydrology and water resources engineering, agriculture, etc. Electromagnetic radiation levels can provide information on vegetation density and species, rock and soil types, water temperature and turbidity, settlement location and intensity, and many other important physical and social variables. Remote sensing makes it possible to view and evaluate conditions in a relatively large spatial area, at a relatively low cost compared to ground surveys.

Cameras and scanning radiometers measure the levels of radiation (visible light or infrared) distributed over some area of interest. The data gathered by these devices can be maintained and presented in either *digital* or *visual* form. In digital form, the measurements are represented by a set of numeric values, one number per *resolution element* or *pixel* (the smallest area from which a measure of radiation can be obtained.) The same measurements, however, can also be considered as a visual *image*. An image is a two-dimensional 'picture' of sensed radiation levels; positions on the image correspond (more or less) to positions on the earth, and the brightness of the image at any position is proportional to the amount of radiation measured at that point.

A visual image is the primary representation provided by a traditional, film-based camera. Special devices (called *scanners* or *frame grabbers*) are necessary to transform a hard-copy photograph into digital form. For data gathered by digital cameras, or aerial or space-based radiometers, the digital form is usually primary.

A variety of computer-based devices can be used to turn this digital data into a visual image. Figure 1.1 illustrates the basic idea behind remote sensing image acquisition and the relationship between visual and digital representations of a remotely sensed image.

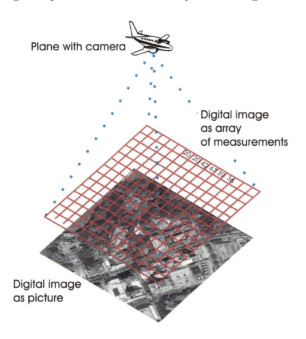

Both digital and visual image representations of remotely sensed data have certain advantages. Visual images are relatively easy for humans to analyze and interpret. The visual representation makes it possible for the human analyst to see spatial patterns in data and recognize objects or areas based on their appearance and spatial relationships. This sort of global relational information is difficult to extract from a set of numbers. However, there is information in a remotely-sensed image that is not directly available to the human perceiver. Distortions, sensor defects and other sources of noise mask some of the information in an image. Important patterns of information exist that can be detected only by combining data from several wavelengths of reflected radiation. Manipulations to minimize noise or reveal multidimensional patterns are difficult or impossible to perform optically or photographically on the visual form of the image. Such manipulations are relatively easy to perform on the digital form of the image data, using a computer-based digital image analysis system.

Figure 1.1: *Remote sensing images: digital and visual*

A *digital image analysis system* is a combination of computer hardware and software that can display, transform and combine image data so that the human analyst can extract as much information as possible. The system maintains and manipulates data in numeric (digital) form, but displays it in visual form. The flexibility and speed of computers makes it possible to try different image combinations, different color assignments, and different noise reduction techniques. The computer can also be used to search for multidimensional patterns not visible to the human eye.

Dragon is an example of a digital image analysis system. It is one of the few such systems designed for use by novices.

1.2   Geographic Information Systems

Geographic information systems, or GIS, are closely related to remote sensing image analysis systems. Like a digital image analysis system, a GIS is a collection of hardware, software, and data used to study and draw conclusions about phenomena distributed in space, usually on the the surface of the earth. GIS systems can be used for applications like forestry, agriculture, disaster planning, and urban planning, just like remote sensing analysis systems. Like remote sensing results, GIS results can be displayed as map-like pictures.

The primary distinction between remote sensing and GIS analysis is that remote sensing works with data that represent more or less raw measurements of earth-related phenomena. GIS in contrast works with data from many sources, which has often been coded, interpreted, grouped, or combined with other information. GIS can handle data *layers* that can be directly observed, such as vegetation type or water bodies, but is just as likely to deal with spatially distributed information about population, income, ecological sensitivity, and location of cases of infectious diseases, information that cannot be 'sensed' in any physical way.

In fact, information derived from remote sensing data is frequently used in geographic information systems: forest or crop cover layers created by classifying satellite images, roads extracted by tracing out their locations with an aerial photo in the background, flood plain maps derived by examining images of flooded areas, slope maps derived from analysis of overlapping stereo images, and so on.

Like remote sensing images, GIS layers can be represented as grids of numbers, with each cell in the grid (each pixel) representing some location in space. **Dragon** has some capabilities for displaying and analyzing GIS data that is represented in this form (called a *raster representation*). Sometimes, however, it is more convenient to store GIS information in a *vector representation*, based on points, lines or regions explicitly encoded by their geographic coordinates. At present, **Dragon** does not provide extensive capabilities for analyzing GIS data in vector form. However, it does allow users to create vector layers with an image as a background (as in the roads example above). It is also possible to import vector data from other sources and display these data as an overlay with a remote sensing image as a background.

GIS makes it possible to combine data of different types in order to answer questions such as:

- Which plots of land are publicly owned, are within 100 meters of a body of water, and have an elevation of less than 200 meters above sea level?
- How is crop yield related to soil type, fertilizer use, and irrigation frequency?
- What regions of the city have expanded and become more dense in the past ten years?
- Is there a relationship between the income from each branch of a store and the median income of the population within five miles of that branch?
- What is the fastest route from each village in a rural area to the nearest hospital?

The answers to these questions can be used to help make decisions such as:

- Where to locate a new state park;
- Where to plant and how to care for the crops;
- How to adjust city zoning laws;
- Where to locate a new branch;
- Where to build or expand rural roads.

A GIS can execute both spatial and logical operations on the data it manages. For instance, it can calculate the proximity (distance), adjacency (connection) and inclusion relationships between different objects or regions. It can also identify conjunctions or disjunctions of properties at a particular location, e.g. whether a building is both in a commercial zone and over six stories tall.

GIS is an important, complex, and fascinating topic. For further information, the reader should consult some of the references in the *Bibliography*.

1.3 Core Image Processing Concepts

Raster Image Representation

Dragon is a program that processes image data. For **Dragon**'s purposes, an image is a two-dimensional array of numbers, each number associated with a position in the array. This kind of numeric array is called a *raster*. Rasters can be used to represent directly sensed information (such as measurements of reflected radiation recorded by a satellite), derived information (such as a categorization of vegetation type produced by analyzing reflected radiation patterns), or synthetic information (such as the reported number of malaria cases at each location).

Positions in the raster array are called *pixels* or *cells*. Each pixel or cell can be identified by a set of *coordinates*. In **Dragon**, the vertical coordinate is usually called the *line number*; the horizontal coordinate is called the *pixel number*. (Both the line number and pixel number coordinates start with a value of zero at the upper left corner of the image.) The maximum possible values of line and pixel number depend on the particular image. The pixel number at a particular point on an image is also sometimes referred to as the point's *X coordinate* while the line number is the point's *Y coordinate*. The pair of numbers is called the point's *image coordinates*.

The values stored in each element of the array are sometimes called *pixel values* or *cell values*. When these values represent measurements of reflected or emitted radiation captured by a sensor, they are sometimes called *digital numbers* or *DN*.

Each cell in a raster normally corresponds to some position on the earth. A variety of different *geographic coordinate systems* exist to describe locations on the earth's surface. Probably the most familiar is *latitude* and *longitude*. Latitude locates a point vertically (in the **y** direction). Longitude locates a point horizontally (in the **x** direction). The latitude/longitude system is a *polar coordinate system* based on angles from the axis of the earth. Lat/long coordinates are measured in degrees. Because the lines of longitude converge at the poles, the size of a degree of longitude varies depending on the corresponding latitude. (See Figure 1.2.) In general, it is not possible to place a grid on a sphere (which the earth approximates) and have all the cells be the same size. Thus, it is not possible to accurately measure distances or areas in terms of lat/long.

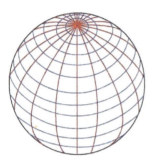

Red lines = Meridiens (longitude)
Blue lines = Parallels (latitude)

Figure 1.2: *The lat/long coordinate system*

Other geographic coordinate systems attempt to remedy this problem by allowing distortions of various types. If one considers only a small area (a few hundred square kilometers, for example), the surface of the earth is very close to being a plane. One popular geographic coordinate system that takes advantage of this fact is the *Universal Transverse Mercator* or *UTM* system. UTM coordinates are expressed as meters rather than degrees. UTM breaks the world into zones and creates a new coordinate system for each zone. Within each zone, the curvature of the earth is minimal, so calculation of distances or areas within a UTM zone are quite accurate.

UTM is one example of a *projection*. Strictly speaking, a projection is a specified mathematical relationship between a two dimensional coordinate system and the actual curved surface of the earth. It is defined by a set of equations plus a specific measurement of the earth's shape. A detailed discussion of projection is beyond the scope of this book. However, readers can think of a projection as a particular recipe for flattening out part of the earth. Readers should recognize that every projection distorts some aspect of space in the process of mapping a curved surface (the earth) onto a plane.

The geometry of a remotely sensed image acquired from space is distorted by various factors. The distance of the sensor from the earth, the angle at which the sensor is aimed, and most important, the curvature of the target (the earth) contribute to this distortion. Hence, it is not immediately possible to relate an x,y position in the image raster to a set of corresponding geographic coordinates. A mathematical process called *georeferencing* or *geometric correction* is required in order to establish this relationship. Intuitively, one can think of geometric correction as a process of "stretching", "shrinking" or "twisting" an image so that it matches the coordinate system provided by a particular projection.

Georeferencing images is one of the first steps in remote sensing analysis. However, some available imagery may already be georeferenced as part of its preparation. Furthermore, modern airborne imagery can sometimes be georeferenced while it is being acquired, using on-board GPS receivers. Once an image has been georeferenced, a particular x,y position in the image can be identified with a specific position on the ground. This means that it is possible to calculate distances and areas more or less accurately.

Multispectral Imagery

One common type image data that can be used with **Dragon** is satellite image data gathered with scanning radiometers such as the Landsat TM or SPOT instrument. These sensors usually measure reflected radiation in several different wavelengths, often called *bands* (or, in some books and image processing systems, *channels*). Thus, a complete set of satellite data for a particular scene will actually consist of several bands, all acquired simultaneously. Such data is called *multispectral*. Multispectral data are useful because some kinds of ground cover reflect very little radiation in some wavelengths and a great deal in others. Thus, a particular ground cover type can be identified by its pattern of radiation levels across different bands (its *spectral signature*). Figure 1.3 illustrates the concept of multispectral image data.

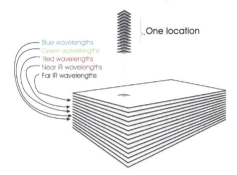

Figure 1.3: *Multispectral data: spatially registered, different wavelengths*

In each band of a multispectral scene, a particular pixel (identified by a specific set of image coordinates) represents the same location on the ground. Because of this property, all the bands of a multispectral scene are said to be *registered*. Registration is an important consideration in digital image processing. It normally does not make sense to combine two images if they are not registered, since you may be combining data from two different locations. Images of the same geographic area acquired at different times usually will not be registered. It is possible to perform a mathematical transformation on one image to register it to another. This transformation is very similar to the calculations needed to georeference an image to a geographic coordinate system.

Image Resolution

As explained above, a raster image is an array or table of numeric values, each of which represents some measurement at a particular location. One important question to ask is 'how large is that location?' or as the question is normally phrased in remote sensing image analysis, 'what is the pixel size?'. The answer depends on many factors including the design of the sensor that acquired the image, the distance of the sensor from the target, and the pre-processing, if any, that has been applied. In general, though, one can state that the smaller the pixel size, the more detail will be visible in the image. Conversely, for a raster of a particular size (for example, 1000 x 1000 pixels), the smaller the pixel size, the smaller the overall area the image will cover.

Figures 1.4 illustrates these principles. The figure shows an area on the ground with a house, a barn, a lake, and a road that runs along the shore. The green area is intended to be vegetation. Two different sensors acquire an image of this area. Sensor A has a pixel size of 20 meters per pixel. Since the house is approximately 60 meters square and the barn 40 meters, each is represented in several pixels in resulting image. The road is about 20 meters wide, so in the resulting image there are a series of road pixels which form a roughly linear feature. The lake is distinct from the road and fills multiple pixels as well.

Sensor B, in contrast, has a pixel size of 60 meters. In the image from this sensor, it is not possible to separate the house and the barn. They have been merged together. The value stored in the corresponding pixel will include measured radiation from both buildings. Similarly, the road is not visible at all. Most of its length is contained in the same pixels as the lake. In the part of the road that is distant from the lake the road disappears because a much larger portion of the area is background vegetation.

The pixel size associated with an image or a sensor is often referred to as its *resolution*. Images (or sensors) with small pixels are called 'high resolution'. Images with large pixels are called 'low resolution'. However, these terms are relative. In the early days of remote sensing (the 1980s), the French SPOT satellite, with 10 meter pixels, were viewed as 'high resolution', compared to Landsat Multispectral Scanner (MSS), which offered 70 meter pixels. Current commercial 'high resolution' imagery, in contrast, provide pixel sizes smaller than one meter, and the resolution is expected to increase.

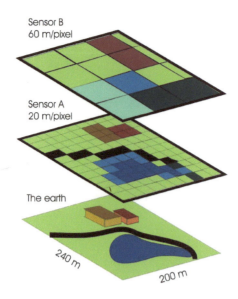

Figure 1.4: *Pixel Size and Image Detail*

One thing that has not changed, though, is the trade-off between resolution and area of coverage. A 1000 x 1000 image from a satellite with 20 meter pixels will cover 400 square kilometers. A 1000 x 1000 image from a satellite with 1 meter pixels covers a single square kilometer. Looked at another way, the higher resolution satellite would require a 20,000 x 20,000 image in order to represent the corresponding are on the ground. Even given the size of modern disks and the speed of modern processors, this much data would stress the capabilities of most remote sensing image analysis systems.

Higher resolution imagery tends to be more visually compelling because it shows much more detail. However, the "best" image resolution depends on the goals of the analysis. For urban planning, where the analyst wants to be able to distinguish individual roads and buildings, imagery with sub-meter pixels is likely to be very useful. To analyze phenomena that are more widely distributed, such as crop health, drought, or deforestation, a larger pixel size will be more appropriate. With sub-meter pixels, an analyst would literally be unable to see the forest because of the trees.

The concept of *scale* is closely related to resolution. Higher resolution images are sometimes described as being "larger scale". The term scale is most often used in the context of a map, where it refers to the relationship between a unit distance on the map and a unit in the real world. For example, a scale of 1:5000 indicates that one centimeter on a map corresponds to a distance of 50 meters in actual space. A scale of 1:5000 is considered to be "larger" than a scale of 1:20,000. A large scale map, as in the case of a high resolution image, shows more detaile than a smaller scale map, but also covers a smaller area assuming a map sheet of the same physical size.

Coded Rasters

The data in remotely sensed images are based on physical measurements. The values associated with pixels are positive integers whose relative magnitudes indicate the intensity of (most commonly) reflected electromagnetic radiation from a particular location. 'Raw' images of this sort can be subjected to arithmetic operations such as ratios or filtering.

Processing raw images can produce a different sort of raster, which **Dragon** calls a *coded image*. A coded image is an array of numeric values. However, the values represent categories or labels, not measurements. Their magnitudes and ordering are arbitrary, and hence it does not make sense to subject such 'images' to arithmetic operations.

Copyright ©Global Software Institute

The most common source of coded images in **Dragon** is the classification process. *Image classification* assigns each pixel position in an image to one of a small number of categories, usually based on patterns of measurements across different bands. For example,the pixels in an image might be identified as "Forest", "Crops", "Wasteland", "Urban" or "Water". Each category would be associated with a value (e.g. 1 for Forest, 2 for Crops, etc.). The value that corresponds to a particular category or class depends on the user's choice, not on the measured data.

Geographic information systems are another source of coded rasters. For example, one might have a raster representing land ownership within a region. The particular numbers used for public versus private land would not be measurements, but simply labels.

Dragon treats coded rasters somewhat differently than images which represent measurements. Coded rasters are a common input to the system's rule-based multi-criteria decision making function.

Vector Data

Dragon primarily operates on raster data, that is, a representation of space as a uniform matrix of rectangular cells or pixels. However, the system has some capabilities to create, import and display vector data.

In a raster representation, the basic unit is a pixel or a cell. In a vector representation, the basic unit is a *feature*. A feature is an abstraction of some object or entity. This can be a physical object on the earth's surface, such as a road, a lake or a cornfield, or a human-defined object, such as a building lot or a province. In either case, the vector representation records the location of the object as an ordered set of points defined in some coordinate system and assumed to be connected.

There are three basic types of vector features: points, lines and polygons. A *point* feature is defined by a single pair or triple of coordinates. It represents a location with no dimensions or extent, for example, the North Pole. A *line* feature is defined by a series of coordinate pairs or triples, assumed to be connected in order. A road or a river might be represented as a line feature. A *polygon* is defined by a series of coordinate pairs or triples that are assumed to be connected. Furthermore, the last point is connected (explicitly or implicitly depending on the system) to the first, so that a polygon forms a closed figure. Line features have length, while polygon features have a perimeter and an area.

Figure 1.5 shows examples of the three basic types of vector features. The light blue is a point (shown as an X to mark its location), the green is a line feature and the purple is a polygon feature. These vector features use an arbitrary coordinate system, but normally vector features will be expressed in terms of a geographic coordinate system such as UTM.

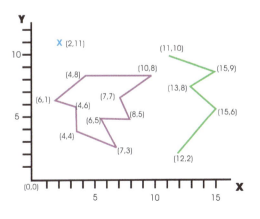

Figure 1.5: *Vector Features*

Rasters represent measurements or categories directly, while encoding spatial location indirectly. In order to know the location of a pixel in a raster, you must calculate that location based on the geographic coordinates of the origin and the size of each pixel in the x and y dimensions.

Vectors, in contrast, represent spatial location directly. On the other hand, measurements or categories associated with a feature must be stored in some sort of auxiliary representation. Each vector feature must have a unique identifier. This identifier can then be linked to various *attributes* or properties of the feature. For example, a line feature that represented a road might have associated attributes such as name, width, surface type, and date built.

Geographic information systems that use a vector representation usually group vector features together into layers, *themes* or *coverages* based on the type of objects they represent. Features representing lakes and rivers, for instance, might be grouped into a "Water" theme, while roads and railroads might be grouped into a "Transportation" theme.

Vector and raster representations offer alternative ways to look at and store information about space. Each representation has advantages for certain kinds of problems and calculations. It is possible to translate from one representation to the other, but the process will almost always introduce errors.

1.4   Remote Sensing Image Analysis Operations

Overview

A remote sensing image analysis system like **Dragon** performs calculations on the cell values in one or more rasters. Sometimes these operations will produce new raster images or layers. In all cases, the user decides what calculations to apply based on her goals.

An analyst's detailed goals, and the corresponding operations she applies, will strongly depend on the specifics of the problem she is trying to solve. However, we can identify some general objectives:

- Visually evaluate or interpret images;
- Manipulate or modify image data in order to highlight certain aspects and make images easier to interpret visually;
- Categorize pixels into classes based on patterns of measured data;
- Explore images numerically or statistically;
- Transform image geometry to match a geographic coordinate system or another image;
- Import, export, and manage data.

Dragon groups operations more or less based on these general goals. The **Display** operations allow the user to examine and compare images visually, as well as to zoom in to inspect details. The **Enhance** operations make it possible to manipulate data from one or more images in ways that highlight (for example) the location of vegetation, emphasize linear features, or suppress noise. The **Classify** operations offer various techniques for categorizing image pixels, based either on examples or on the intrinsic structure of the image data. The **Geography** branch of the system provides facilities for geometric correction, image measurement and statistical analysis, as well as operations for creating and displaying vector features. The **File** and **Utility** branches provide necessary data exploration and management capabilities.

An Example Scenario

A simple list of **Dragon**'s capabilities probably does not help a reader new to remote sensing to understand the overall process. This section presents a simple scenario that illustrates how some of these operations might interact.

Thailand experiences distinct wet and dry seasons. During the dry season, normally from February through April, little if any rain falls. As a result, drought is a frequent problem.

We would like to compare the surface area of the Lamtaklong Reservoir (located in Nakhon Ratchasima province, northeastern Thailand) between the wet and dry season. We have two images captured by the Enhanced Thematic Mapper sensor on Landsat 5. The pixel size is 25 meters. We have three bands, blue, green and red, for each scene. Figure 1.6 shows *natural color composite* displays of the wet season and dry season (left and right images,respectively). The difference in the size of the reservoir is easy to see, but we would like to evaluate it in a quantitative way.

Normally, our first step would be to georeference both sets of bands. However, these images were georeferenced by the Geoinformatics and Space Technology Development Agency (GISTDA), the organization that supplied them. Both images have been corrected to correspond with the UTM coordinate system. Thus, the two images from different times are registered to one another.

We want to isolate the water in the two different images so that we can compare them. One way to do this is to *classify* the images, assigning each pixel to a category based on the values of its three bands. We decided to use a *supervised classification* procedure. In this method, the user must select examples of each class she

Figure 1.6: *Wet and Dry Season Three Band Composite - Imagery Courtesy of GISTDA*

wants to identify. This is called *training* the classification. Figure 1.7 illustrates part of the training process. The user has identified regions of the image and labeled them as "Water". The user will presumably also identify at least one class of non-water as well. The software then uses these as examples to find pixels with similar values in the three bands. (This shows only one of several training methods available in **Dragon**.)

Figure 1.7: *Selecting Training Areas for Water Class*

The training process calculates a *signature* for each class, a set of statistics that describe the selected examples. We then run a classification procedure to label the other pixels in the image according to one of our classes. **Dragon** offers several supervised classification methods. In this scenario, we use *maximum likelihood classification*. The results from classifying these two data sets are shown in Figure 1.8.

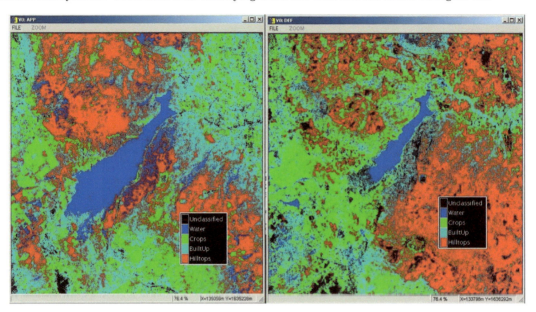

Figure 1.8: *Classification Results for Wet and Dry Season Data*

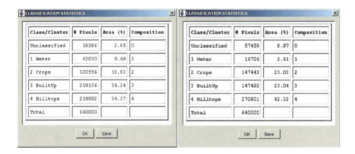

Figure 1.9: *Classification Statistics for Wet and Dry Seasons*

The classification process displays a table showing the number of pixels assigned to each class. Figure 1.9 shows the numerical results for the wet and the dry season classifications. These results allow us to calculate the difference in water area between the two image dates. In the wet season image, 62030 pixels, or nearly 10% of the image, is classified as water. In the dry season, only 16706 pixels, less than 3% of the area, is water. Since each pixel in this image is 25 meters square (or 625 square meters), the difference in number of pixels (45324) corresponds to an area of more than 28 million square meters, or about 28 square kilometers.

Note

These results probably overestimate the difference. For this example, we selected training areas without any actual information about what is on the ground at each point. The classified images show "water pixels" in areas that are clearly not water. The results are intended only to show a typical sequence of operations in a remote sensing project and should not be treated as authoritative.

We might want to visualize this difference in the form of an image. To do so, we can use the **Combine Layers** operation. This operation allows us to create rules which will assign new values in the output based on combinations of values in multiple input images. We create a rule which says to set the output value to 127 (an arbitrary non-zero value) if the value of the first image (the wet season classification) is 1 (water) and the value in the second image (the dry season classification) is anything except water. All other pixel positions will be set to zero. The results of this conditional recoding are shown in Figure 1.10. The difference in the size of the reservoir is dramatic. However, note that there are other water areas shown as well. These are probably errors due to inaccuracies in the original classification results, as discussed above.

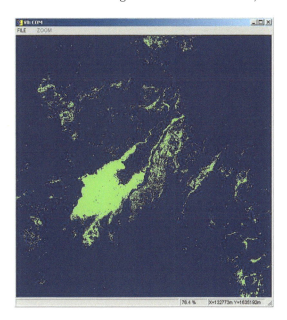

Figure 1.10: *Conditional Recoding to Show Reservoir Changes*

The scenario above is fairly typical. The details will vary depending on the problem and the objectives. Regardless of the problem, however, remote sensing image analysis usually involves a series of operations, each of which uses and transforms the results of previous operations. Some operations (like classification) involve arithmetic calculations. Others (like the conditional recoding) are based on logic. Often the analysis will require the user to interact with or visually interpret the image (as in training).

Dragon provides the basic operations needed to solve a wide variety of problems involving remotely sensed imagery and other geospatial data. This book will help you understand these operations and how to use them appropriately.

Chapter 2

INTRODUCTION TO DRAGON

2.1   What is Dragon and What Can It Do?

Dragon/ips® and **OpenDragon**™ are software packages for personal computers that display and manipulate images, i.e. visual representations of data. (In the remainder of this book, we will refer to either of these packages as **Dragon** unless we need to be specific.) Although **Dragon** can be used for a variety of different types of visual information, it is oriented primarily toward processing remotely-sensed images.

This chapter provides an overview of the functions available in **Dragon**. It also describes the **Dragon** user interface and the various ways that you can interact with the system in order to analyze remote sensing and GIS data.

Dragon Image Processing Capabilities

This section summarizes the image processing capabilities available with **Dragon**. It assumes that you have read Chapter 1 and have some familiarity with basic remote sensing concepts. For more information about these capabilities, consult the chapters on individual **Dragon** operations and the resources listed in the *Bibliography*.

Dragon has capabilities for displaying single-band images in color or gray-scale form. It also includes three-band color composite and image overlay capabilities. **Dragon** provides facilities for adding titles or labels to screen images and for displaying legends for classified images. The user can explicitly control contrast stretching and color assignment or allow **Dragon** to handle these processes automatically.

Many common enhancement operations are available in **Dragon**, including band sums, ratios and differences, edge enhancement, smoothing and arbitrary 3x3 filters. A few more unusual enhancement capabilities are also provided: symmetric differencing, relaxation, and masking. The results of each enhancement operation remain available without explicitly saving them to disk, so that multi-step computations such as vegetation indices can be accomplished efficiently. Images created by any **Dragon** operation or imported from outside sources can also serve as input to enhancement processes. **Dragon** also includes transformations such as principal components analysis as part of its image enhancement capabilities.

Dragon supports density slicing, supervised classification using boxcar (parallelepiped), minimum distance to mean, and maximum likelihood algorithms, and unsupervised clustering. Training samples are selected interactively and can be saved, examined, combined, and modified. Training area boundaries can also be saved and then applied to a new image. This allows identical areas to be used in two separate classifications (e.g. in multitemporal studies). The resulting image can be modified by combining or recoding classes. **Dragon** provides control over colors assigned to classes. Classified images can be visually overlaid on the original image or combined with the original using the masking option to assist in the interpretation of class contents.

Dragon also includes facilities for geometric correction and registration, and for point, line, and polygon operations. The geometric correction operations consist of interactive ground control point selection, calculation of image transformation equations, and resampling of the image. Point, line, and polygon operations include provisions for creating vector feature overlays, for extracting polygonal subimages ('cookie cutter' operations), and for measuring polygons and polylines.

Dragon also offers facilities for basic raster GIS modeling. Capabilities include calculating slope and aspect images from digital elevation model input data; identifying pixels that belong to a buffer within a certain distance of target pixels; and logically combining multiple raster layers based on a set of conditional rules.

Finally, **Dragon** offers a variety of utility functions including: image data value display, image subsetting, histograms, scatterplots, display and modification of identifying information in image file headers, and saving images.

A variety of stand-alone programs are included with **Dragon**. These include programs for exchanging data with other remote sensing and GIS systems.

The basic capabilities of **Dragon** can be augmented by means of special-purpose routines created by the individual user. This possibility is discussed in more detail in the the manuals supplied with the **Dragon** Programmer's Toolkit or the **libdtk** libraries.

2.2 Dragon Documentation

The *User Manual* is the primary documentation for the **Dragon** software system. The complete and most up-to-date version of the *User Manual* is provided in machine readable form both as a Portable Document Format (PDF) file,and as a set of HTML files. A reader program supplied with **Dragon** provides convenient access to the HTML version of the manual. You can also view the individual files using any web browser such as Firefox, Opera, or Internet Explorer.

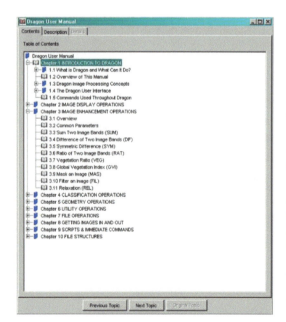

Figure 2.1: *Manual reader table of contents pane*

Dragon includes an application specialized for reading the *User Manual*. This application, called the Manual Reader, is an enhanced HTML browser that knows about the structure of the **Dragon** documentation. The Manual Reader is used to provide context-sensitive operation-specific help within Dragon. However, it can also be used in a stand-alone mode to browse the manual as a whole.

To bring up the Manual Reader when you are running **Dragon**, choose the **Help⇒User Manual** menu item.

The Manual Reader has three tabbed panes. The leftmost tab, labeled **Contents**, displays a table of contents in a tree-structured form. Figure 2.1 shows the Manual Reader contents page. When you select a topic from the tree, the center tab, labeled **Description**, comes to the front, showing the text of the currently-selected topic. Buttons at the bottom of the application allow you to navigate to the next or previous topic in the document.

The third tab, labeled **Details**, is enabled only for sections of the manual that describe specific **Dragon** operations. In these cases, clicking on **Details** will display a table showing all the parameters associated with the operation, with their meanings, defaults, and the **Dragon** command language code (*parameter specifier*) for each one. If the topic displayed in the **Description** pane does not relate to a specific **Dragon** operation, the **Details** tab will be disabled. Figure 2.2 shows a sample **Details** page.

When the **Details** pane is displayed, the **Previous Topic** and **Next Topic** buttons are disabled. You must return to one of the other panes to select a new topic.

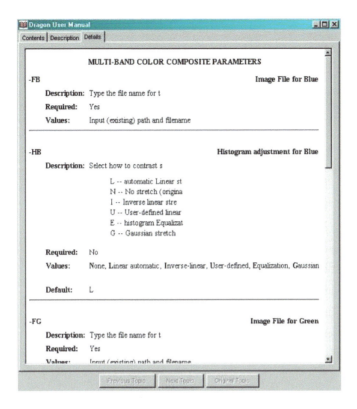

Figure 2.2: *Manual reader details pane*

As noted above, the Manual Reader can be accessed from inside **Dragon** by choosing **Help**⇒**User Manual**. If the Manual Reader is not already visible, it will be activated. Then, if **Dragon** is currently displaying the data entry screen (*response panel*) for a particular **Dragon** operation, the Manual Reader will immediately display the **Description** pane for that operation. Otherwise, the Manual Reader will simply display the first page of the first chapter.

2.3 Dragon Image Processing Concepts

Images and Image Files

Dragon is a program that processes image data. As described in Chapter 1, an image is a two-dimensional array of numbers, each number associated with a position in the array. Positions in this array are called *pixels* or *cells*. The number of cells in a remote sensing image can be very large: up to 16,000 lines by 16,000 columns or 256 million pixels. Each pixel may be contained in multiple bands of data (e.g. 8 bands for ETM, although ETM images are less than 16,000 by 16,000). Most remote sensing data has radiance or coded values less with less than 256 levels and can be stored as one 8-bit byte per per pixel per band. However, there are exceptions: NOAA AVHRR data has 12 bits, and *digital elevation model (DEM)* data are frequently stored as 16 bits.

Image data in the **Dragon** system are stored in *image files*. The image files contain the data values for each pixel, line by line. Image files are normal operating system files: you can copy them, rename them, delete them, etc.

Image files in **Dragon** can have any name up to 32 characters long (without any embedded spaces). However, all image files must have the extension (filetype) **.img** (or **.IMG**) so that **Dragon** can recognize them. Normally, the name you give to an image file should tell you something about what the image data is and where it came from. For example, you might call a file holding aerial photography data from Bangkok **BangkokPhoto.img**.

As discussed in Chapter 1, many satellites measure reflected radiation in multiple different wavelengths, often called *bands*. This sort of multispectral data is often essential for identifying phenomena on the ground. Two types of land cover may look very similar in one spectral band, but quite different in another. In order to distinguish the two, multiple bands are needed. For example, both water and vegetation may reflect light in the green region of the electromagnetic spectrum. However, water reflects very little energy in the infrared region, while vegetation strongly reflects infrared. Thus, analyzing both infrared and green bands makes it possible to discriminate between these two categories.

In some image processing systems, data from all the bands of one scene are stored together in the same file. In **Dragon**, each band of a scene is stored in a separate image file, so you have complete control over which image bands to use. An inconvenience of this is that you must explicitly decide which bands to include or exclude in any operation.

Since you frequently want to process several bands of the same scene together, it is important to know not only which scene but also which band the data in a file represents. We recommend that you use a standard method for naming your image files, such as reserving the last character in the file name for the band number. For files that represent the same scene, all other characters in the name would be the same. Thus, **Bangkok1.img**, **Bangkok2.img** and **Bangkok4.img** would be appropriate names for three bands of a multispectral image of Bangkok.

Coded Image Files

A standard image file is assumed by **Dragon** to be *radiometric data* derived from a sensor. The numbers which represent each image element are therefore physical measurements of the emitted or reflected radiation received by the sensor, and arithmetic operations (such as histogram normalization) can be reasonably applied to these measurements.

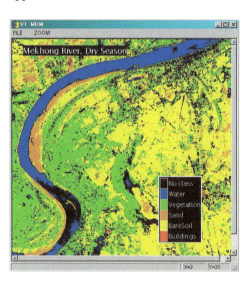

Figure 2.3: *A coded image, with legend*

As discused in *Chapter 1*, some files have the same structure as radiometric image files, but the numbers are codes or category labels rather than true physical measurements. They may carry order information but never magnitude information. For this reason, it does not make much sense to perform arithmetic operations on such coded images. **Dragon** refers to these files as *coded image* files, and processes them slightly differently than radiometric image files.

Structurally, a coded image is identical to any other image: a two dimensional array of numeric values, where each pixel value is associated with some position on the ground. **Dragon** will not stop you from performing arithmetic on coded files, but it will warn you that you are doing something unusual.

Coded image files can arise in several different ways. For example, a classified image results from applying a classification (categorization) procedure to one or more image bands. A classified image is distinguished by having a C in its header *File Type* field, and by having associated class names. Figure 2.3 shows an example of a classified image, including a legend that shows the association of each class with its display color.

When **Dragon** displays a classified image, it will not perform any default contrast stretch (since that would change the class values). Also, a classified image has its own associated color scheme, created when you save the image after performing a classification. This special color scheme, which can be modified using the **Utility**⇒**Colors** operation, is always used to display the classified image in color unless you explicitly request a different color scheme.

Copyright ©Global Software Institute

A coded overlay image can also be generated using the **Geography**⇒**Vector** operations, and a coded layer file may be imported from a Geographic Information System (GIS) package. The **Geography**⇒**Aspect** and **Geography**⇒**Combine Layers** operations also produce coded layer files. Layer files are characterized by having an *L* (for *Layer*) in the header *File Type* field, and by including somewhat different ancillary information than a classified file.

Dragon uses the *File Type* field in the image header to determine whether an image is coded or not. If you want to perform some unusual operation such as a contrast stretch on a classified image, you can do so by changing the value of the *File Type* from *C* to *I*. This will 'fool' **Dragon** into treating the file like a normal image. Likewise you can make a normal image look like a classified image to **Dragon** by changing its *File Type* to *C*. **Dragon** permits you to do almost any sort of bizarre processing, but tries to protect you from doing so by accident.

Image Header Information

In addition to the data values for each pixel, every **Dragon** image file contains a block of identifying information called the *header*. The information in this header is used to keep track of where image files came from, what they represent, and what processing they have received. The general term for this type of descriptive information is *metadata*, that is, data about data.

The different items of information stored in the image file header are called *header fields*. **Dragon** provides you with capabilities for examining header fields and for changing many of them. (See the chapter on Utility operations.) The most important header fields that you can list and/or modify are:

Identifying Fields

- **Number of Lines:** how many lines are in the image represented by this file (Cannot be changed)
- **Number of Pixels:** how long each line is, in pixels (Cannot be changed)
- **File Type:** *I* if a normal or *radiometric* image file, *C* if a classified image *L* if a GIS layer or rasterized vector file.
- **Scene Identification:** a text label that describes the primary location from which the data was recorded (should be the same for all bands of a given scene)
- **Subscene Identification:** a second text label that identifies a subset of the primary location from which the data was recorded (should also be the same for all bands of a given scene)
- **Band Number:** a label, usually a number, that identifies the band (if any) the image represents
- **Source:** a three-character code identifying the **Dragon** operation that created the file
- **Comment:** a phrase with any additional identifying information you wish to enter
- **Color File Name:** optional name of color scheme file to be used by default in displaying the image.

Statistical Fields (These cannot be directly changed by the user.)

- **Mean value:** arithmetic average of data in file, to nearest whole number
- **Maximum value:** largest data value in file (range 0 to 65535)
- **Minimum value:** smallest data value in file (range 0 to 65535)
- **Variance value:** a measure of the 'spread' of values around the mean
- **Standard Deviation:** another measure of spread
- **Histogram:** table of frequencies counting how many pixels in the image have each possible value

Special Fields

- **Names for each class:** labels for classes or specific data values in the image
- **Georeferencing information:** fields that define the relationship of the image coordinates to an external coordinate system.
- **Calibration information:** fields that define a mapping between image data values to some external measured quantity, such as temperature or elevation.

Dragon recalculates image statistics after any operation that could change image values, and generally tries to maintain the header information in the computer memory in a correct state. **Dragon** uses the identifying fields in the header for checking logical consistency in multi-band operations, and the statistical fields for computing automatic contrast stretch parameters, density slicing boundaries, etc.

The Main Memory Image

Dragon's operation depends on the concept of a *main memory image*. All operations on images that produce a single band result are considered to take place in the main memory image area, and all results are left in the main memory image. Consequently the main memory image usually holds the results of the most recent computation. This is a central concept in the organization of **Dragon**.

The main memory image is very convenient because it lets you chain together a series of operations without having to write intermediate results out into files on your disk. Suppose, for example, you wanted to add together two image bands, and then subtract a third from the result. You would choose the operation for adding image bands and tell **Dragon** what image files held the data. **Dragon** would read the first set of data into the main memory image, then read and add successive lines from the second set of data. The result of the addition would be left in that memory image. Next you would choose the operation for subtracting one image from another. This time, when **Dragon** asked you what files held your data, you would select the current image in memory (referred to by the 'special filename' =**M**) for the first set of data. There would be no need to save the results of the addition in an image file and then read that data back into memory for the subtraction. Memory images make the sequence of operations quite efficient.

Mask Images: Processing Subregions of Images

Many of the **Dragon** analysis and classification operations can be limited to specified subregions of the image data. These regions, which are defined by a *mask image*, can be arbitrarily complex in shape and contain numerous disconnected regions. In general, you can think of the mask image being overlaid on the image data. Wherever the mask image is zero, the image data value is ignored and the resulting image will have zero value. Where the mask image is not zero, computation proceeds as usual. Figure 2.4 illustrates the concept of using a mask.

Because the mask data need not have the same dimensions (number of lines and pixels) as the image data, an arbitrary choice must be made about image data *outside* the mask image area. In most operations, pixels which are completely outside the mask image are treated as though the mask were **not** zero.

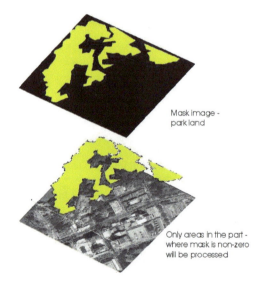

Mask image -
park land

Only areas in the part -
where mask is non-zero
will be processed

Figure 2.4: *Mask images can restrict processing to relevant areas*

The **Classify⇒Recode** operation uses the mask image slightly differently. In this operation, pixel positions where the mask image is non-zero, as well as any which are outside the area of the mask image, will be recoded. Positions where the mask image is defined and is zero are left unchanged.

The **Utility**⇒**Histogram** operation also allows you to specify a mask image. If you do specify a mask, the histogram calculations include only the pixels where the corresponding mask image pixel is non-zero. Among its other uses, this function allows you to see a histogram of the raw data values of only those pixels which were classified as members of one particular class. For example, to see a histogram of the water pixels in your image, create a mask image using **Classify**⇒**Recode** based on your classified image with all non-water pixels recoded to zero. Then specify that mask as the mask image in **Utility**⇒**Histogram**.

All of the operations which permit use of a mask image (generally, most of the **Enhancement** and **Classification** operations) provide a field to specify a mask image file name. If this field is left blank, no mask will be applied.

Color Schemes and Color Files

A *color scheme*, in **Dragon** terminology, is an association between each of the 256 possible image data values (0 to 255) of a coded image and the color in which pixels with that value should be displayed. A color scheme may assign the same color to more than one data value. Frequently a contiguous range of values will be assigned to the same color, for example, all values between 16 and 31 might be assigned to light-green. You can, however, assign the same color to dissimilar image data values. This sort of color scheme is frequently used for special effects.

Color schemes are stored in *color files* until they are used. Color files can have any name up to eight characters long in **OpenDragon**, or up to thirty two characters long in **Dragon Professional**, but there must be no embedded spaces. The file extension (filetype) must be **.clf** (or **.CLF**). You should not create or modify any color files with names of the form **defxxxxx.clf**. These are standard color files that **Dragon** uses in various operations. The standard color files are located in a subdirectory within the system directory, called **defcolor**.

Dragon includes an operation for defining and modifying color schemes and saving them in color files Thus, you can have control over the color assignments in your images. On the other hand, **Dragon** has a set of standard color schemes that often provide good results. You do not have to define your own color schemes unless you want to achieve some particular effect not possible with the standard schemes.

Using **Utility**⇒**Header**, you can store the name of a color file in the header of an image file. Then **Dragon** will use that color scheme, by default, whenever you display the image. Classified images always include the name of a color file in their headers, as discussed in above. Note that **Dragon** assumes that a color file recorded in an image file header is located in the same subdirectory as the image itself.

Other Dragon Files

There are several other kinds of data files used in **Dragon**. They are summarized here; more detailed descriptions plus information on the structure of these files, and other files used by **Dragon**, are included the *User Manual*.

Files Associated with Classification

- *Signature files* (extension **.sig**) store *training signature statistics*, values that are used to guide the classification process by defining typical values for each class. Signature files are created by the **Classify**⇒**Edit Signatures**⇒**Save** signature operation.
- *Polygon files* (extension **.ply**) record the boundaries of training areas selected in the **Classify**⇒**Training** operations.
- A *recoding table file* may be used in combination with the **Classify**⇒**Recode** operation to provide an arbitrary remapping of image data values. Refer to the *User Manual* for details.

Files Associated with Geometry

- *Vector files* (extension **.vec**) are used to capture the coordinates and attributes of points, lines and polygons created in **Geography**⇒**Vector**. The captured data can be exported to other software packages which use vector data, or they can be read by **Geography**⇒**Vector** to become the basis for creating additional vectors, or to be transformed into a raster data layer.
- *Measurement data file* (no standard extension) are used to capture the data from the line profile and area histogram displays in the **Geography**⇒**Measure** operation. These output data files allow you to do statistical or graphical analyses on numerical values gathered from **Dragon**, using spreadsheet, statistics, data base, or presentation graphics software. The file format for the output data file is a generic comma-delimited form that is intended to be easy to import into other packages.
- *Coefficient files* (extension **.cof**) record the regression coefficients calculated by **Geography**⇒**Calculate**.
- *Ground control point files* (extension **.gcp**) record coordinates and user-defined labels for ground control points. GCP files can hold either image or map coordinates. They are created as an output of the **Geography**⇒**GCP** operation, and serve as input to **Geography**⇒**Calculate**.

Miscellaneous Files

- *Script files* (extension **.spt**) are text files of **Dragon** commands, created with any word processor or editor. **Dragon**'s script capability makes it possible to execute almost any series of commands automatically, without any user interaction.
- *Rule files* (extension **.rul**) are used to specify how to assign result values in the **Geography**⇒**Combine Layers** operation. and can be generated using the **Geography**⇒**Create Rules** operation or by hand.

Filenames and Special Filenames

Selecting a File by Name

There are many times while using **Dragon**, when the user must specify a filename. As described below, a *File chooser* is almost always available to assist. Simply press **<F3>** to get a list of choices in the current *default directory*. If it is not in the current directory, you can navigate to other directories.

You may, if you wish, provide the full *drive:\path\nodename* form in order to access files which would not otherwise be available. Usually, you do not need to specify the file extension.

Special Filenames and the Status Window

Whenever **Dragon** permits you to specify the name of an image file to read, you are being asked which data the software should operate on. Frequently that data is stored in a file on your hard disk, or on your network. However, **Dragon** also permits you to operate on data already 'in memory'.

Dragon provides several *memory images* which you can use as input in some operations. You tell **Dragon** to use one of these as the source for your operations through *special filenames*. A special file name is a sequence of characters which would be illegal as a real file name. The most important ones are listed below.

- The **=M** (the 'M' stands for the English word *main*) memory image contains the results of the most recent image analysis operation. All **Enhance** operations and all **Classification** and **Geography** operations which produce an image result place that result into **=M**. Because **=M** can be used as the (or an) input to most of these operations, it is very easy and fast in **Dragon** to chain a sequence of operations on the same data. In addition to the analysis operations, the **Display**⇒**1 Band**, **Display**⇒**Gray**, and **Display**⇒**Overlay** operations, when reading from a disk file, also load that file into **=M**. Note that the main memory image is an 8-bit image.
- The **=R** (*Red*) memory image contains the image which was assigned to the red color plane during the most recent **Display**⇒**3 Band** operation. **=R** can be the input to most display and many analysis operations.

- The =**G** (*Green*) memory image contains the image which was assigned to the green color plane during the most recent **Display⇒3 Band** operation.
- The =**B** (*Blue*) memory image contains the image which was assigned to the blue color plane during the most recent **Display⇒3 Band** operation.
- The special filename =**C** does not correspond to a single memory image, but rather to the collection =**R**, =**G**, and =**B**. =**C** (the 'C' stands for the English word *composite*) provides a convenient way to use a multi-band image as a reference or background for operations such as **Utility⇒Cursor** and **Geography⇒GCP**. Note that it does not make any sense to specify =**C** as the input to an analysis operation.

In all cases, the memory images contain the true data values with no stretch applied. Thus, it is very fast and easy to experiment with viewing the effects of the different stretch algorithms supplied. For those operations which permit you to specify a composite image (=**C**), the composite image is displayed using the same stretches which were most recently used for **Display⇒3 Band**.

There may be times, however, when you want to change the image values by applying a histogram stretch. **Classify⇒Recode** provides some very powerful capabilities for doing this. The results of such a stretch can be saved as a new image using **File⇒Save**.

Not every operation or image file name field will accept every special filename. Generally the limitations are reasonable or obvious; however, the Help message for each field tells you which ones are permitted.

Of course, you also cannot specify a name of a special file if the corresponding memory image has no image loaded into it. To help you to remember which images are loaded, **Dragon** provides a *Status display* listing the memory files as well as other information. To see the Status window, press <**F6**> or click on **Status** radio button with your mouse. The Status window alternates with the History display. You can switch to the History display by pressing <**F7**> or selecting the **History** radio button.

How Dragon Finds a File

There are two types of files which **Dragon** needs to be able to locate:

- **Dragon** *system files*, a category which includes the executables, help files, configuration files, and various other kinds of files.
- *Data files*, including all image, script, signature, vector, polygon, and registration files, plus most color files.

To assist you in organizing and accessing your hard disk, **Dragon** provides a facility for a *default image path* or *default data directory*. This is a drive and path specification where **Dragon** will automatically look for any input file which you may name.

For example, if the default data directory is **C:\images** and you type, in answer to some question requesting an image file name, a file name of **bangphoto**, **Dragon** will look for the file

```
C:\images\bangphoto.img
```

If the file is not actually located in the **C:\images** directory, your will receive an error message saying the file cannot be found. If, in fact, the file is on some other directory (for example, on **C:\landsat**), you would have to specifically name the directory by typing

```
C:\landsat\bangphoto.img
```

You must specify a value for the default data directory during the installation of the **Dragon** software. The default data directory can be changed while **Dragon** is executing by use of the **File⇒Preferences⇒Paths** command.

There is also a default location, called the *default file creation path*, where **Dragon** will create new data files. Like the default image path, the initial value for this path must be specified during the installation process, but can be changed via the **Preferences** submenu. Frequently it is useful to have these two paths refer to the same location, so that you will automatically be able read files that you write, without changing the directory specification.

We have been discussing how to type a file name in order to identify a file for **Dragon**. In fact, a user rarely needs to type file names, since it is always possible to browse and select files graphically by using the **Dragon** file chooser. To bring up the file chooser, click on the down arrow button at the end of a file name data entry field, or press **<F3>**.

A few **Dragon** operations allow you to select multiple files in one step. The selected files can be typed, in which case they are added to a list when you press Enter, or they can be selected in a file chooser. Once the list of files has been created, the user can change their order or delete specific items using appropriately labeled buttons.

2.4   The Dragon User Interface

The **Dragon** interface consists of the following components:

The Dragon menu client This is a window which presents a set of menus for selecting operations within **Dragon**. When a menu item is selected, typically the menu client will display a data entry form, called in **Dragon** terminology a *response panel*, where you can supply the parameters necessary to the operation. The menu client also provides a window for displaying image status information or command history, an optional command line, and several status areas at the bottom of the window. The status area on the left indicates whether **Dragon** is currently working on an operation or is ready for input. The larger status area on the right displays information messages, such as the number of pixels truncated in the Filtering operation or changed in the Relaxation operation.

Help When your mouse enters the boundary of a menu item, or a control on a response panel, **Dragon** will wait for a short interval and then pop up a *Help window* near the control. The help window provides you with information on the meaning and use of the current control, in the currently selected natural language. To make the help window disappear, you can either move the mouse away from the control, or click in the control. You can change the delay between the time that your mouse enters a control area and the time that the help appears by editing the **HELPDELAY** value in the **dragon.cfg** configuration file.

Viewports All images are displayed in special *Viewport* windows. **Dragon** provides two viewports, labeled Viewport 0 and Viewport 1. Most image processing operations allow you to select which viewport to use for displaying the results. Viewports have their own menus, which control image specific operations such as zooming and printing images. A status line at the bottom of each viewport provides information such as the current cursor (mouse) position and brief instructions about the current operation.

InfoViewports *InfoViewport* windows are small sub-windows used for displaying ancillary graphics, such as histograms or scatterplots. They can be iconized or closed when not needed. Multiple InfoViewports may be visible simultaneously.

Report Windows A number of **Dragon** operations are available to allow you to view information about images, signatures and so on. This sort of information is displayed in a *report window*. The results tables created by classification operations are also shown in report windows. Like InfoViewports, report windows can be dismissed when you are finished with them. Most also offer the opportunity to save their contents as as HTML file.

Button Panels Some of the interactive operations in **Dragon**, such as **Classify⇒Training**, display a floating panel of push buttons which can be used to invoke particular actions. The buttons on these panels are enabled appropriately depending on the state of the interaction. These buttons always have function key equivalents to invoke their functionality.

Query Dialogs Some of the interactive operations in **Dragon** also display simple query dialogs. For instance, if you click on the **New Class** button during **Classify⇒Training**, **Dragon** will display a dialog where you can enter or select the next class to train.

File Chooser Dragon provides an enhanced file chooser dialog that makes it easy for you to locate and select the files you want to use. The file chooser provides directory tree and file list areas typical of Windows file selection dialogs, but also offers some advanced capabilities including a set of selectable recently used files and the ability to filter the visible list of files by typing a partial file name into the

Copyright ©Global Software Institute

File Pattern field. To display the file chooser, click on the downward pointing arrow at the end of an input field that expects a file, or press <**F3**>.

Error and Message Boxes As is the case with most Windows applications, **Dragon** uses pop-up message boxes to communicate the existence of error or warning conditions. Usually these boxes have a single button, labeled **OK**, for you to dismiss them. Occasionally, they will require that you make a choice (e.g. **Exit** versus **Continue**).

Figure 2.5 shows many of the important aspects of the **Dragon** Menu Client.

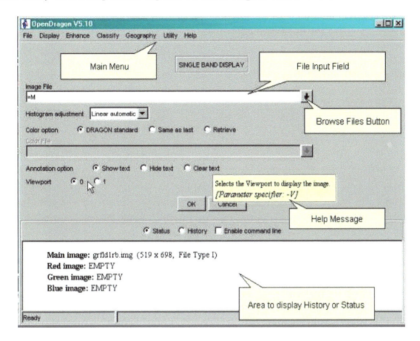

Figure 2.5: *User interface components*

Interacting with the Viewports

As explained above, **Dragon** provides two special windows called viewports which are used to display images. Each viewport has its own menu bar and some specialized functionality, which is discussed in this section.

Most **Dragon** image processing operations finish by displaying an image which holds the results of the operation. Usually you can decide whether the results should be shown in Viewport 0 or Viewport 1.

Interaction with a viewport is the place where differences between **OpenDragon** and **Dragon Professional** are most obvious.

Dragon Professional can process images of (almost) unlimited size. For the current generation of satellites, a *full scene* image can contain up to about 16,000 rows and columns of data. Processing this much data is not a problem for **Dragon**, but the result will have about 400 times more data than can appear on a typical computer screen. This gives rise to two conflicting requirements: you want to see the full scene results, but you also want to see (and work with) the full resolution contained in your data. **Dragon Professional** lets you do either of these, but not both at the same time.

In **Dragon Professional**, each time you display a new image, whether from disk or as the result of some image processing, the entire image will be displayed but in a reduced resolution. We call this the *overview* image. The actual scaling factor, the ratio between the displayed data and the full image, is calculated based on your screen size, viewport size, and the full image size. This scaling factor is shown in the status line below the viewport.

In order to see more detail in part of the image, you can select a *region of interest* by positioning the mouse at one corner of the desired region. Press and hold the left button while dragging the mouse to the opposite corner of the desired region. During this time, a white rectangle will outline the current region and the mouse coordinates will be displayed on the right size of the viewport status line. Normally, if the image is georeferenced the coordinates will the the actual UTM values.

After you select a region of interest, you can further refine your selection by choosing a new region within the current region using the mouse in the same way. Alternatively, you can return to the overview by clicking the *right* mouse button.

There may be times when you begin to select a region but change your mind. There are two ways to cancel a selection in process:

1. If the selection you make is too small (less than about 200 image rows x 200 image columns), **Dragon** reject the selection. Thus, if you simply press and then immediately release the mouse button no region will be selected.

2. Alternatively, if you begin a region selection and then move the mouse outside of the viewport, that will also cancel the selection.

In addition, you can never select a region such that the display scale will be larger than 100%, which is one data element per screen pixel.

This region selection process works well when you simply want to view your processed images. However, there are several operations where your need to *interact* with the image by selecting particular points, possibly individual image data points, for classification training, image registration, etc. These additional requirements make the above process somewhat more complex.

The six (at present) interactive operations fall into two groups:

- Operations where the desired points **must** be chosen at maximum resolution. These operations are classification training, ground control selection, and **Utility⇒Cursor**. These operations are almost meaningless unless you are selecting individual data points. Therefore, you must select a region of interest before proceeding with the point selections. You can return to the overview and select a new region as often as you wish.

  The requirement that your region of interest must have full resolution does however, restrict you to selecting a region that will fit on your screen.

- The other interactive operations are the **Geography⇒Polygon** operations. These operations permit you to draw vectors either in the overview image or in the detailed 1:1 or 100% region image. Because mouse clicks in the overview image are used for drawing, you must use a specific command (**<F11>**) to begin the region selection.

In **OpenDragon**, image results are always displayed on the screen at a 1:1 ratio (100%). That is, one pixel on the screen is exactly one cell of image data. Images slightly too large for the screen are displayed with vertical and/or horizontal scrollbars as needed. Likewise, if the viewport is resized to be smaller than the current images, scrollbars will appear. This is a good arrangement for teaching, because students do not need to deal with complexities of scaling. All interactive operation involving selecting points in the image exactly correspond to selecting a single pixel on the screen. However, scrollbars in both X and Y directions do not work well for very large images. It is easy to become lost, and not know what part of an image is being shown or how to find other parts of the image. Thus, **OpenDragon** can only handle images of limited size.

OpenDragon will normally size the viewport conveniently given the size of the monitor screen. However, you can manually resize the viewport to make it larger or smaller. If necessary, scrollbars will appear, which will allow you to view the entire image contents, even when the viewport is quite small.

There are several **Dragon** operations where you must move a graphics cursor around in the viewport. These include training area definition, coordinate display, color scheme assignment, and most Geography operations. There are two ways you can control the movement of the cursor that represents your 'current' position.

The easiest way is to use a mouse. **Dragon** can use almost any mouse or trackball for cursor movement, and the mouse can be used interchangeably with the keyboard. **Dragon** does expect that your mouse will have at least two buttons (referred to in this manual as the left and right mouse buttons).

Instead of using a mouse, you can use the *arrow keys* available on the keyboard of most personal computers. There are usually four arrow keys, pointing up, down, left and right. Each time you press an arrow key, the cursor will move in the corresponding direction.

Once you have the mouse in the desired position, you can select a point by clicking the left mouse button, or by pressing the <**Ins**> key on the keypad.

The **Zoom** menu allows you to control the degree of magnification of the image during interactive operations.

- Magnify 3x
- Enlarge
- Restore
- Overview

The first two items enlarge the image, so that you can see more detail. To do a 3x zoom of the image, select the **Magnify 3x** item, then use the mouse to select the location on the image that you want to be in the center of the zoomed image. To zoom to an arbitrary expansion factor (up to 10x), select **Enlarge**. **Dragon** displays a dialog for you to enter the desired zoom factor. After typing a number between 1.0 and 10.0 and clicking the **OK** button, click on the image to select the center of the zoom area.

Choose the third item, **Restore**, to redisplay the image region at full resolution (1:1 scale).

The fourth item reduces the image until it fits in the current size of the viewport. In contrast to the previous two menu options, **Overview** does not require any interaction. As soon as you select the menu item, the image will be shrunk to fit the viewport size and displayed.

After you zoom in, the viewport disables the two 'zoom' menu items listed above and enables the **Restore** option. Thus, you cannot do multiple successive zooms. If you enlarge the image to an arbitrary factor, and want to zoom in further, you need to restore the image to normal resolution first, and then re-execute the **Enlarge** option using a different magnification factor.

In **Dragon Professional**, the zoom capability is enabled only when the viewport is displaying a full resolution, 1:1 scale region. Furthermore, the **Overview** item redisplays the entire image at reduced resolution.

The viewports provide facilities for printing the current image, saving all or part of the image to a file, or copying all or part of the image to the clipboard. All these capabilities are accessible through the **File** menu on the viewport.

To print the currently-displayed image, choose the viewport **File⇒Print** menu item. **Dragon** will display a standard Windows print dialog, allowing you to select a printer where the image data should be sent. Obviously, your printer must have color capabilities if you want to correctly represent a color image. Almost any printer will do a reasonably good job with a gray-scale image.

To save the current image in TIFF format, a format that can be imported into most graphics and word-processing programs, choose **File⇒Capture**. The viewport will display a file selection dialog so that you can browse to the location where you want to save the image, and enter a filename. After you click on the **Ok** button on the file dialog, you decide what to save. If you want to save the entire viewport, press <**F1**>. Otherwise, place the mouse cursor on the image at the upper left corner of the area that you want to capture. Press the left mouse button, and drag the mouse to the lower right corner of the area you want to select. A flexible box will follow the mouse. When you release the mouse button, the box will disappear. The viewport will convert the image contents to TIFF format and save them in the designated file.

To capture all or part of the image to the Windows clipboard, choose **File⇒Copy**. Once again, you can copy the entire viewport to the clipboard by pressing the <**F1**>, or use the mouse cursor to choose a subarea. The viewport will copy the all the data or just the section included within the selection box to the clipboard. From there you can past the data into graphics programs or word processing documents.

To cancel a copy or capture operation, drag the mouse outside the viewport area.

At any time that there is an image displayed in a viewport, you can find out the X (pixel) and Y (line) coordinates of any point in that image. As you move the mouse cursor over the image, two fields at the bottom right of the viewport display the continuously updated coordinates of the cursor.

Note

In **Dragon Professional**, you can control whether the mouse position display uses geographic coordinates (if available) or image coordinates, by editing the file **dragon.cfg**, in the **Dragon** installation directory.

The viewport only accepts point selections (clicks) at very specific times in the processing sequence. If you try to select a point on the image when the viewport is not accepting points, the viewport will beep and ignore the click.

Interaction Modes

Dragon's user interface design accommodates varying levels of skill and experience by providing two quite different modes of operation within the same program. For users new to **Dragon**, to remote sensing image processing, or to computers, the program offers the *Menu* mode of operation. **Dragon** menus gives detailed and explicit information and instructions about each step needed to accomplish a task. It prompts for each piece of information that you must supply and gives immediate feedback if you make an error. Menu operations are discussed in the following section.

The Menu mode offers considerable assistance to a new user. However, experienced users may get impatient if they must go through prompts in order to accomplish an operation. For these users, **Dragon** provides the *Command* mode. In Command mode, the program does not prompt for specific information. You indicate exactly what you want by typing in a command that identifies an image processing operation plus command *parameters* that further specify the action: the source of data, the color scheme to be used, etc.

In addition to the Menu and Command modes, you can also use **Dragon** in *Script mode*. Script files hold **Dragon** commands, in exactly the same format you would use in Command mode.

Menu Mode

The **Dragon** menu can be presented in a variety of languages other than English. You can change the language used by accessing **File⇒Preferences⇒Language**. (The new language will be shown the next time you start **Dragon**.) When **Dragon** is configured for a language other than English, all menus, prompts, messages, and help will be displayed using that language, in the appropriate character set.

When working in Menu mode, **Dragon** uses *menus*, which are lists of items to select, and *response panels*, which are data entry forms to fill out.

Dragon menus work like those in any other Windows application. Click on the menu to drop it down and display its items. Use the mouse to click on the item you want, or use the arrow keys to navigate to that item and hit the space bar to select it.

Response panels act like standard Windows data entry forms, but have a few additional characteristics.

- Pressing the **<Enter>** key is equivalent to pressing **<Tab>**; it simply causes the focus to move to the next field. It does not activate the **Ok** button, as is the case in many programs. This allows you to cycle through the data fields multiple times before triggering the operation.

- Pressing the down-arrow key is equivalent to pressing **<Tab>**; it causes the focus to move to the next field. Pressing the up-arrow key is equivalent to **<Shift-Tab>**, moving backwards through the fields on the form.

- Pressing the **<F3>** key in a field where a filename is required will bring up a file dialog so you can browse to the file. You can accomplish the same thing with the mouse by clicking on the down-arrow graphic at the end of the field.

- Pressing the **<F3>** key in a field represented by a drop down list control (also called a 'combo box') causes the choice list to drop down.

- Pressing the **<Home>** key in a field that holds some data clears the field. Left and right arrow keys move the insertion cursor within the field.

- Pressing the **<End>** key in a field that holds some data moves the insertion cursor to the end of that data.

- In some cases, the information that you enter in one field will cause changes in other fields when you tab off the original field. The most common change is to enable or disable other fields.

- To indicate that you are finished entering parameters and want to actually execute the operation, click on **Ok** or press the **<F1>** key. To cancel the operation, click on **Cancel**, or press either the **<F10>** or **<Esc>** keys

The bottom section of the menu client is a multi-function text area. This section can be used to display image status information, or command history. You choose which category of information to display by clicking on the appropriate radio buttons, or by pressing function keys: **<F6>** for Status or **<F7>** for history.

When the Status functionality is active, **Dragon** displays a summary of what image information corresponds to the special files =**M**, =**B**, =**G**, and =**R**. This includes the image file names, if the data have not been modified, and the image dimensions.

When the History functionality is active, **Dragon** displays a list of all the commands that have been executed, in the **Dragon** command language discussed in the next section. Even if you do not have History selected, the history information will continue to accumulate. Thus you can switch among the different information modes (*Status* and *History*), without losing any information.

In additional to the information mode radio buttons, the bar between the response panel area and information display area contains a check box labeled *Enable command line*. This check box displays or hides the command line; pressing the function key **<F2>** is equivalent. The use of the command line is discussed briefly in the next section, and in more detail in the *User Manual*.

Command-Line Mode

The Menu and Command modes of interaction look quite different. However, the logical structure of operations and the specifying information (*parameters*) each operation requires are almost exactly parallel in the two modes. You can move easily from one mode to the other, with a single key. Thus, you have great freedom in how much of each mode you use. As you learn more about **Dragon**, you can start to use the Command mode for some familiar operations while still using Menu mode for more unfamiliar ones. The choice is entirely yours.

In **Dragon** Command mode, you use the same menu choices, operations, parameters and parameters values as in Menu mode. You just express them somewhat differently.

All information that you type in Command mode is typed on the *command line*. To make the command line appear, click the check box labeled *Enable command line*, or press **<F2>** when a response panel is visible. The command line is a single line text entry area between the response panel area and the status/history area. It is labeled *CMD>>*.

When the command line is visible and active, **Dragon** still allows you to operate the menus. If you select a new operation from the menu, the command line will disappear and you will be returned to Menu mode. On the other hand, if there is a response panel displayed when you turn on the command line, the panel will remain on the screen, with all fields disabled. Even though the fields are disabled, you can still view the help messages for each field. This can assist you as you work to construct a **Dragon** command in the command line area.

For more detailed information about the command-line mode, refer to the *User Manual*. Detailed arguments for the command line operations are available in the Manual Reader.

Script Mode

A **Dragon** *script* is a normal text file that you create with a text editor program. Script files hold **Dragon** commands, in exactly the same format you would use in Command mode. A script can contain a single command or a long series of commands.

To actually execute a script, choose **File⇒Script⇒Run** and navigate to the desired script.

When you click **OK**, **Dragon** begins reading and executing the commands in the selected script file. It continues executing commands until it reaches the end of the file (unless you choose to cancel the script execution part way through).

Scripts can be used for many purposes: to automate a series of operations that are performed frequently, to create 'packaged' command sequences for limiting the typing required from novice users, and to create demonstrations and interactive tutorials. For details, refer to the *User Manual*.

2.5   Commands Used Throughout Dragon

There are a few commands or control keys that can be used in all operations or sub-menus within **Dragon**. Some of these commands have been discussed before in other contexts, but all are included here for for reference.

Special Keys

Computer keyboards typically include a set of special keys called *function keys* that do not exist on a typewriter keyboard. **Dragon** uses many of these special keys to trigger frequent types of action or as synonyms for frequently used commands.

PC keyboards also include a keypad that contains other special keys. These keys include the **<arrow>** keys (up, down, left, and right), the **<Home>** key, **<PgUp>** (*Page Up*), and **<PgDn>** (*Page Down*) keys. Some of the keypad keys are also used for **Dragon** actions, as explained below.

Commands for Operation Control

Several control key sequences are used for special purposes when you have selected an operation and are entering parameter values. These control sequences are effective in either Menu or Command mode.

- Press **<F1>** to begin execution of the current operation. This key sequence is also used to accept the current values and dismiss a sub-panel.

- Press **<F2>** to switch from Menu to Command mode or vice-versa when a response panel is displayed.

- Press **<F3>** or click on the downward pointing triangle in a response panel field to get a list of possible values for that field, or to bring up a file selection dialog, for filename fields.

- Press **<F6>** to switch the multifunction text area in a response panel so that it displays the current system status information.

- Press **<F7>** to switch the multifunction text area in a response panel so that it displays history of recent commands and messages.

- Press **<F10>** or **<Esc>** to cancel the current operation and remove the current response panel from view. These key sequences are also used to cancel and dismiss a sub-panel.

Chapter 3

IMAGE DISPLAY OPERATIONS

3.1 Principles of Image Display

In order to visually interpret remotely sensed images, you must display them on the monitor screen. What you will see depends on the data stored in image files, which could be digital satellite or aerial image data, aerial photography data that has been converted to digital form using a scanner, classification results, or non-image geographic data such as elevation or population.

In any kind of image display, the numeric pixel values in the image file are associated with colors or levels of gray. This allows a human analyst to see spatial patterns in the values. If the image was captured by measuring reflected radiation from the earth, larger values in the image file will correspond to brighter areas on the ground. Assigning darker shades of gray to smaller values and brighter shades to larger values will result in an image that "looks like" what the sensor saw. Figure 3.1 provides a simple illustration of this principle. This kind of display is called *monochrome* or *gray scale* display.

Array of pixel values

Each value assigned
to a different gray level

Figure 3.1: *Basic principle of raster image display*

As explained in Chapter 1, many sensors simultaneously measure and record reflected radiation in multiple different spectral bands. A common combination found on many satellites measures a blue or blue-green band (roughly 400-500 nm wavelength), a green band (500-600 nm), a red band (600-700 nm), and one or more infrared bands (near infrared 700-900 nm, mid infrared 1500-1700, etc.). The wavelength boundaries for different bands, as well as the number of bands, vary from sensor to sensor.

When a single band of a multispectral (that is, multi-band) image is displayed in shades of gray, areas of high reflectance will appear bright and areas of low reflectance will appear dark. Since some objects on the ground reflect strongly in one band but absorb radiation in other bands, the patterns of brightness can look quite different from one band to another. For example, vegetation reflects green wavelengths (which is why trees and grass appear green) and near-infrared. but absorbs almost all red light. Thus, forested areas will show up as relatively bright areas in a gray scale display of the green band while they will appear dark in the corresponding red or infrared band. See Figure 3.2 for an illustration.

Almost any color can be created by combining the primary colors (blue, green and red) in different intensities. In particular, computer monitors create color displays by varying the amount of blue, green and red light they output at each point on the screen. A combination of equal amounts of blue and green, with no red, will produce cyan (bluish green). A combination of green and red, with no blue, results in yellow. Blue and red without green produces magenta (a purplish hue). Varying the levels of different components will change the resulting color accordingly. Equal levels of all three components produces a shade of gray ranging from black (all three components at minimum level) to white (all components at maximum level).

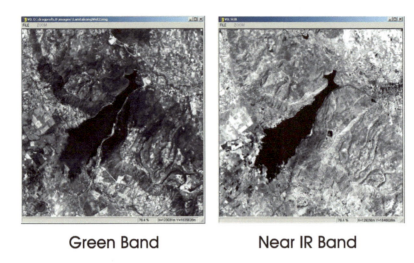

Green Band   **Near IR Band**

Figure 3.2: *Gray scale images of different spectral bands*

Multispectral remote sensing images offer the opportunity to create color displays by using one band to determine blue intensities, one band to determine green, and one band to determine red. This type of display is called a *three-band composite* display. Three-band composites are extremely useful for visual analysis of remotely sensed imagery.

One common way to create a composite is to assign the blue band of a remote sensing image to the blue component of the display, the green band to the green component, and the red band to the red component. (Of course this is possible only for sensors that offer blue, green and red bands.) This type of display is called a *true color* or *natural color composite* because it looks approximately like what the human eye would see. Areas that look blue on the ground look blue in the composite, and so on. Figure 3.3 shows a natural color composite made up of the blue, green and red bands from the image used in Figure 3.2.

Figure 3.3: *Natural color composite image*

Frequently, however, it is useful to assign bands to display components in other combinations, in order to highlight specific information. In particular, informative displays can result when image bands from outside the visible range of the spectrum (e.g. infra-red bands) are included in a composite. Any composite image in which one of more bands are assigned to components that do not match the natural wavelength is called a *false color composite*

One widely-used band combination for false color composites assigns a green band to the blue component of the display, a red band to the green component of the display and a near or mid infra-red band to the red component. In this sort of composite, vegetated areas appear bright red, and clear water becomes nearly black. This display combination is popular partly because it mimics the appearance of images acquired by using color-infrared photography, a technology that was widespread before the advent of satellite and airborne digital sensors. Figures 3.4 and 3.5 show examples of different false color composites.

So far we have been discussing various ways to display radiometric images, that is, raster files where the pixel values represent actual measurements. Gray scale and composite displays are less useful for coded images, that is, images where pixel values represent different categories instead of measured values. In radiometric images, pixels carry magnitude information, so it makes sense to make the brightness proportional to the numeric

Figure 3.4: *False color composite: red band as blue component, blue band as green component, green band as red component*

Figure 3.5: *False color composite similar to color-IR film*

pixel value. (Note that some raster data that do not represent directly sensed information also carry magnitude information, e.g. an elevation layer or a layer showing accumulated rainfall). This is not true of coded images. A value of 10 is not necessarily larger than a value of 9, and the difference between 9 and 10 is not smaller than between 2 and 10.

Coded images are usually visualized using *pseudo-color displays*. A pseudo-color display assigns a different color (blue/green/red combination) to each distinct value in the image file. Pixels that are close in value will not necessarily be close in color. In fact, the color mappings for pseudo-color displays are frequently designed to maximize the visual difference between categories.

Classified images are the most common example of coded images in **Dragon**. A classified image results from running a supervised or unsupervised classification operation. The input is normally a set of remotely sensed image bands. The output is an image in which each pixel is assigned a value based on the calculated category of land cover. A value of 1 might be assigned to forest, 2 to water, 3 to bare soil, and so on. **Dragon** allows you to decide what color to use for displaying each class and to save the mapping of colors to classes in a *color scheme* file for reuse. Figure 3.6 shows an example of a pseudo-color display for a classified image. The legend, which is optional, shows the relationship between colors and classes.

Pseudo-color displays can also be used with radiometric images to achieve special effects. For example, you could create a color scheme in which most values were assigned to shades of gray except for values in a narrow range of interest. These special values could be mapped to red. When you displayed an image in pseudo-color using this file, pixels with values in the range of interest would be highly visible against the gray background.

The next section provides an overview of the operations available in the **Dragon** display menu. Following this, you will find detailed descriptions of each operation. The chapter concludes with a set of hands-on exercises to give you practice displaying and viewing images.

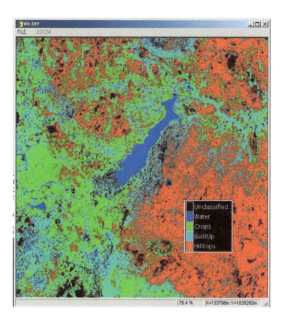

Figure 3.6: *Pseudo-color Display of Classification Results*

3.2   Overview of Dragon Image Display

The **Display** operations include:

- Display of a single data file (single band image) with different data values shown in different colors
- Display of a single band image with different data values shown in different shades of gray
- Combine three different image bands to form a color composite image, in which one band is represented in shades of blue, one in shades of green, and one in shades of red
- Overlay a second image 'on top of' the last-displayed image. The previous image values will 'show through' at any point where the overlaying image pixel has a value of zero.
- Annotation utilities including:
  - Add text operation, which provides a way to label or title images on the screen.
  - Legend operation, which adds a key to a classified image display to indicate the identity of specific classes or data values and the colors assigned to them.
  - Erase text, which provides various possibilities for deleting, erasing, and replacing previously-placed annotations.

3.3   Image Display Characteristics

Dragon requires a computer with a *display resolution* of 1024 or more pixels wide by 768 or more lines high, and a full-color (at least 15 color bits) capability. Most modern computers provide 24-bit color and displays are available up to about 1900 pixels wide. This is still far less than the size of a full-scene TM image.

OpenDragon and **Dragon Professional** handle this problem differently. **OpenDragon** simply limits the size of image you can process to the size you can see. **Dragon Professional**, by contrast, provides advanced capabilities to allow you to work with images of almost any size.

Note that the same image data can be displayed using either the color or gray-scale operations. However, the images displayed through these two operations will appear quite different because a gray-scale display maps data values to colors in an ordered progression, from dark to light, while a color display lacks this ordering.

The range of available colors in color display operations includes several shades of gray. Thus, you can create color schemes for use with color display operations, which in fact hold only gray hues. One such color scheme (**defgra.clf**) is provided with **Dragon**. You should use this color scheme when you want to display a gray background image with colored vectors or a color image superimposed on it.

3.4 Common Parameters

Display operations have some common parameters, which are explained below.

Histogram adjustment

Most **Dragon** display operations allow you to choose the type of histogram adjustment to apply during display. Choices available include several standardized stretch calculations, as well as a user-specified linear stretch. The standardized stretch calculations are *Linear, None, Inverse linear, Equalization,* and *Gaussian stretches.* Figure 3.7 shows the results of several historgram adjustment options.

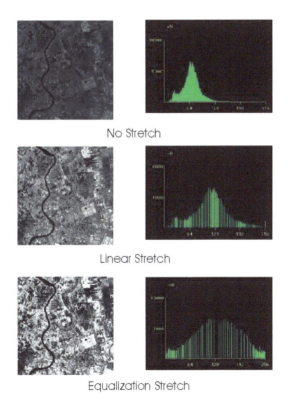

Figure 3.7: *Results of histogram stretching*

If you specify *User-defined,* **Dragon** will display a sub-panel with fields where you specify the upper and lower break points for the stretch. This panel will not appear until you change focus from the *Histogram adjustment* field to a different field, after selecting *User-defined.*

Note that histogram adjustment does not alter the image data at all, only the way it is displayed.

Color option

This option gives you the opportunity to control which color file will be used to display your image. Usually, you will use the **Dragon** default (which is **def1ba.clf** for continuous-value image files, unless a different default color file is specified within the image file). Alternatively, you may choose *Retrieve* to select any color file available on your system, or *Same as last* to reuse the color file most recently used without having to remember what it was.

Annotation operation

As you work with an image, you may use the **Display⇒Annotation** operations to add text and/or a legend to your image. Those operations give you fine-grained control over what appears on your image, and where it is placed. However, you may sometimes wish to see the image without all of your annotation, but without erasing it. At other times, you may wish to get rid of all of the annotation permanently.

The *Annotation option* parameter permits you to decide, each time you re-display an image, whether to *Show* the existing annotation, to *Hide* it, or to *Clear* (erase) it.

Viewport

Almost all **Dragon** display and analysis operations permit you to display the result in one of two possible viewports. (By contrast, most *interactive* operations such as classification training and ground control point selection do not provide a choice of viewports.)

3.5 Single Band Color Display (1BA)

Choose **Display⇒1 Band** to display a single image band in color, that is, to create a pseudo-color display. Different colors will be assigned to different ranges of data values in the image. You can specify a color scheme to use, or use the standard scheme. You can allow **Dragon** to determine an automatic contrast stretch based on image statistics, choose stretch parameters yourself, or specify no stretch.

If you redisplay an image for which you have previously defined text labels and/or a legend, you can show or hide these annotations depending on your choice for the *Annotation option* parameter.

3.6 Three Band Composite Display (3BA)

Choose **Display⇒3 Band** to combine three different bands of data into a single image, assigning one band to shades of blue, one to green, and one to red. Normally, the three image bands will represent light reflected at different wavelengths from the same scene. The contributions from the three bands will be 'blended' for each pixel, and the resulting image will show a variety of colors depending on the relative values of red, green and blue data for each pixel position. **Dragon** normally determines the best contrast stretch for each band, in order to include the maximum information in the final image displayed. However, if you want to explicitly control the stretch for each band, you can do so using the usual *Histogram adjustment* parameters.

3.7 Single Band Gray Image Display (GRA)

Choose **Display⇒Gray** to display a single image band in shades of gray. Different gray shades ranging from black to white will be assigned to different ranges of data values in the image. The standard set of gray shades is the best gray level set that can be produced with your graphics hardware. Thus, **Gray** does not include parameters for selecting your own color scheme. You can also design your own gray scale color scheme by selecting gray shades from the default color scheme used in **1BA**. However, this type of user-defined gray scheme must be used in **1BA**, not in **GRA**.

Gray provides the same options for contrast stretching and annotation display as **Single Band Color**.

Copyright ©Global Software Institute

3.8   Overlay Image (OVE)

Display⇒**Overlay** allows you to overlay a second image 'on top of' the image that you displayed most recently in the selected viewport. The second image will cover the first, except that any pixels in the second image that would normally be displayed in black (that is, which have image data values of zero) will appear 'transparent'. You will be able to see the information from the first, underlying, image, wherever the second image has zero data values. Note that when you choose **Display**⇒**Overlay**, there *must* be some previously-displayed image in the selected viewport.

Overlay is useful for interpreting classified images. If a classified image is overlaid on the original from which it was derived, the original image will 'show through' wherever there are unclassified pixels (since unclassified pixels normally have a value of zero). You can use **Enhance**⇒**Masking** and **Classify**⇒**Recode** to construct special 'mask' images in which classes of interest have the value zero, so the original image values for pixels of those classes can be seen through the overlay. **Overlay** can also be used to display a layer file created with **Geography**⇒**Vector** on top of a background image.

Overlay provides the usual contrast stretch, color file, and viewport options for the overlaid image. The usual color scheme selection options are also provided. However, the underlying image will always be shown using the color scheme in which it was previously displayed.

In **Dragon Professional**, once you have displayed an overlaid image, the normal capabilities for selecting subregions or returning to an overview will be disabled. To see the combined image at a different scale, redisplay the background image and adjust to the desired scale, then repeat the overlay operation.

3.9   Annotate Image (ANN)

Choose **Display**⇒**Annotate** to add text to the current image. Text can serve as titles, labels for particular features in an image, etc. You can also erase or replace text strings. Annotate also lets you place a legend on an image. The legend identifies each class or labeled data value by name and by color.

All of the text you define using these operations, together with the fonts, the screen positions, and the position of a legend, make up the current *annotation set*. The *Annotation option* field in most display operations allows you to control whether the annotation set is visible or not. The current annotation set will be automatically cleared if you display a different image.

Add Text To Image (ADD)

Choose **Display**⇒**Annotation**⇒**Add** to place a line of text on a displayed image. By default, this operation does not affect any previously-specified text. However, you can explicitly choose to replace the previously added text. You can choose any TrueType font, size, and style available on you computer, as shown in the selection dialog associated with the *Text Style* parameter. (The Text Style dialog is shown in Figure 3.8.) The default font is Lucida Sans Unicode, which has the great advantage that is can display all Unicode characters for almost all of the world's languages. You will find that most other fonts on your computer will not have this advantage. (Lucida Sans Unicode is part of the Java system supplied with **Dragon**).

If you change the font, size or style for one annotation operation, **Dragon** assumes that you want to continue to use these parameters for future annotation. To avoid this, uncheck the 'Set selected as default' check box before clicking on **<Ok>**.

Add also requires you to specify a starting position for the text, either interactively using the cursor or in line and pixel coordinates. If you leave the *Line* and *Pixel* parameters blank, you will have the opportunity to place the text interactively, using the mouse. (The parameters refer to the image coordinates of the lower-left corner of the text.)

Add give you the additional choices of which viewport to receive the annotation, and whether this text should replace or be added to the previously-existing text. All text is displayed as white. You can choose to have either an *opaque* (i.e. black) or *transparent* background for the text.

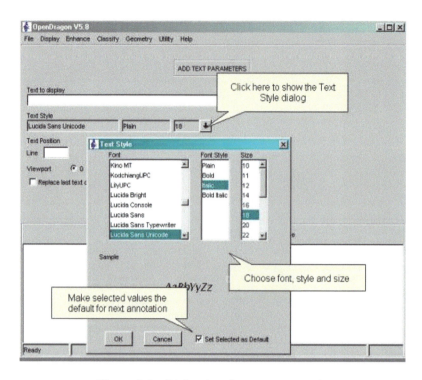

Figure 3.8: *Dialog for choosing text style*

Text to be displayed can include any printable characters including blanks. The printable characters may include non-English characters provided that you have an appropriate font installed on your computer.

Display Classified Image Legend (LEG)

Choose **Display⇒Annotation⇒Legend** to put a legend identifying colors with class names or labels on an image. **Legend** requires you to specify a starting position for the text, either interactively using the cursor or by specifying a line and pixel number (for the lower-left corner of the legend). Leave the *Legend Position* parameter blank if you want to position the legend interactively.

Legend give you the additional choices of which viewport to receive the annotation, and which font should be used for the text portion of the legend. All text is displayed as white on a black background.

The content of the legend is controlled by information in the image header. To change the labels and associated values for an image, use **Utility⇒Header**. To modify the color keys displayed in an legend, you must change the color scheme associated with the image, using **Utility⇒Color** to create a new color file; then either display the image explicitly specifying the new color file, or else use **Utility⇒Header** to modify the default color file in the image file header.

Erase Last Text Display (ERA)

Choose **Display⇒Annotation⇒Erase** to remove previously-displayed text from the image. This operation can be used to delete all text, the most recent text line, or a legend display from the selected viewport. It cannot remove more than one line of text (in a single operation) without removing all text from that viewport.

You do not need to use this command if you merely want to revise or move the most recent text line or legend you displayed. **Add** provides an option to replace the last-added line of text, while re-executing the **Legend** operation for an image automatically removes any previously-placed legend.

Erase can be used repeatedly to remove several annotations from the image, in order of decreasing recency.

3.10   Exercises

This section offers a set of hands-on exercises with **Dragon** display operations, in order to let you practice using these functions and see their results.

These exercises and those in subsequent chapters assume that you have downloaded imagery archive discussed in the *Preface* from the **OpenDragon** website and unpacked its contents into your default data directory. Furthermore, they assume that your default output directory is the same as your default input directly. Thus, the exercises never specify any paths for input or output files, but assume that **Dragon** will supply the default values.

1. Start your **Dragon** system. Choose **Display**⇒**Gray** to display a single band image in shades of gray. Enter the following parameters:

   - Image File: losan1.img
   - Histogram adjustment: None
   - Viewport: 0

   Click the **OK** button to start the operation. **Dragon** reads the image file (the blue band of a Landsat Enhanced Thematic Mapper image, showing part of Los Angeles) and displays it in shades of gray. You can see some of the details like roads, but overall the image is rather dark and hard to analyze.

2. Choose **Display**⇒**Gray** again. Enter the following parameters:

   - =M (This is the default value and means "the image in memory" or the last image processed.)
   - Histogram adjustment: Linear default
   - Viewport: 1

   When you click the **OK**, **Dragon** displays the same image in the other viewport window, using a linear contrast stretch. Notice how much more clearly you can see the details in the new image, compared to the first one you displayed, particularly the bright, built-up areas in the lower right and upper left.

   Histogram adjustment, also known as *contrast adjustment* or *stretching* works by spreading out the values in the image so that they cover the entire range of possible gray values. If you look at the original image in Viewport 0, you will see that most pixels are a medium gray. There are few black or white pixels. Linear histogram adjustment displays the smallest pixel values in the image in black, the largest in white, and distributes the intermediate pixels across the available range of gray values.

   Experiment with the other histogram adjustment options. In particular, you might want to compare histogram equalization with the linear default option. What is the difference between the two resulting images?

3. Choose **Display**⇒**3 Band** to display a color composite image. Enter the following parameters then click on **OK**.

   - Image File for Blue: losan1.img
   - Histogram adjustment for Blue: Gaussian
   - Image File for Green: losan2.img
   - Histogram adjustment for Green: Gaussian
   - Image File for Red: losan3.img
   - Histogram adjustment for Read: Gaussian
   - Viewport: 0

   The resulting image is a natural color composite. The hills, which in Los Angeles' arid climate are mostly bare, show up in shades of brown with tan areas that are cleared regions or quarries. You can see a few areas of green vegetation near the pass, where the highway cuts through the hills. If you zoom in, you'll see a deep blue reservoir near the middle of the image. You can also see the individual rectangular building lots in the urban areas. The resolution of this image is 25 meters per pixel, so larger buildings and major roads show up very clearly.

37

4. Choose **Display⇒3 Band** again. Enter the following parameters then click on **OK**.

   - Image File for Blue: losan2.img
   - Histogram adjustment for Blue: Gaussian
   - Image File for Green: losan3.img
   - Histogram adjustment for Green: Gaussian
   - Image File for Red: losan4.img
   - Histogram adjustment for Read: Gaussian
   - Viewport: 1

The resulting display is a false color composite, where green, red and near infrared image bands are assigned to the blue, green and red display components. The main advantage of this display is that the vegetated areas show up far more clearly than in the natural color composite. Even the hills have some reddish quality, suggesting that they are covered with sparse brush rather than completely bare. The reservoir is also easier to identify in this presentation. The false color composite also shows another body of water in the hills to the left of the pass, which is not visible at all in the true color image.

5. Choose **Display⇒1 Band** to display a pseudo-color image. Enter the following parameters then click on **OK**.

   - Image File: SanFranMaxLikelihood.img
   - Viewport: 0

Dragon displays a classified image, with each class in a different color. To see the image from which this coded image was derived, choose **Display⇒3 Band** and display bands **SanFr1.img**, **SanFr2.img** and **SanFr3.img** as blue, green and red respectively. Display the composite in Viewport 1 so you can continue to examine the pseudo-color image.

6. Select **Display⇒Legend** to display a legend so you can see which colors represent which classes. Accept all of the default parameter values and click on **OK**. **Dragon** will display a cross-shaped cursor on the classified image in Viewport 0. Move the cursor to the location (bottom left corner) where you want the legend to appear and then click the mouse to place the legend.

Compare the classified image colors to what you can see in the natural color three band composite. Notice, for example, the rectangle of Golden Gate Park (a mixture of sparse and dense vegetation classes) and the dense vegetation surrounding Lake Merced south of the park.

7. Select **Display⇒Add** to add a title to the classified image. Enter the following parameters then click **OK**.

   - Text to display: San Francisco Classification
   - Text Style: Lucida Sans Unicode, Italic, 20 point (To select the text style, click on the arrow icon at the end of the field)
   - Text Position Line: 100
   - Text Position Pixel: 20
   - Viewport: 0

The annotation text appears near the upper right corner of the image. If you are using **Dragon Professional** and have selected a sub-region to display, you may have to return to displaying the full image (press the **<Esc>** to do this) in order to see the text.

8. Select **Display⇒1 Band** operation. Enter the following parameters then click **OK**.

   - Image File: =M
   - Color option: Retrieve
   - Color file: mekdmdm.clf
   - Annotation option: Show
   - Viewport: 0

Dragon re-displays the classified image using a different color mapping. Because you chose "Show" for the annotation option, **Dragon** continues to display the legend and the title that you added. Notice that the colors in the legend change to correspond to the color scheme in the newly selected color file.

Chapter 4

IMAGE ENHANCEMENT OPERATIONS

4.1 Principles of Image Enhancement

Image enhancement operations perform transformations on data stored in image files, producing new images with new characteristics. Enhancement operations are frequently used to improve the quality and interpretability of images. In general, they either emphasize certain types of relevant information in the image (e.g. linear features) or minimize the visibility of distracting or irrelevant information (e.g. bright spots caused by sensor defects).

Some of these operations are sometimes also referred to *image arithmetic* operations because they perform common arithmetic operations on pairs of iamges.

Images captured by remote sensors (satellites, aerial cameras and so on) frequently contain errors and distortions. As discussed in *Chapter 1*, some of these errors are geometric. The image does not accurately reflect the positions of objects on the earth. Geometric problems can be at least partly remedied by geometric correction, which is discussed in Chapter 6. However, remotely sensed images also may contain various radiometric errors that make the image difficult to interpret. For example, the pixel values in some images may be restricted to a fairly narrow range, resulting in an image with poor contrast (as illustrated in the Exercises in the previous chapter). An image might contain systematic noise, such as a repeating pattern of blurred bands, or random noise, such as bright or dark dots scattered through the picture. Any of these problems will interfere with your ability to extract useful information from the image.

Image enhancement is the process of modifying an image in order to improve its appearance or its interpretability. Images can be enhanced using a wide variety of different methods. You have already seen how histogram adjustment can improve image contrast and hence your ability to see details in an image. All image enhancement methods operate by performing various calculations on the numeric pixel values.

Image enhancement is useful not only for reducing the impact of radiometric errors and noise, but also for emphasizing or revealing specific kinds of image information. For example, *edge enhancement* operations make linear features in an image (roads, rivers, building outlines) more visible. Sometimes combining two image bands arithmetically can reveal previously hidden information. *Vegetation index* calculations use the ratio of two bands to highlight areas with high chlorophyll content. Subtraction of one band from another (usually bands from two different images that have been registered) can be used to identify areas of change.

The **Dragon** enhancement menu offers a wide variety of different operations for improving image interpretability. Some of these operations, particularly those that combine bands, can also be used for raster GIS operations. For example, two raster layers that represent scores for ecological diversity and distance from the nearest road could be added together using **Enhance⇒Sum**. The pixels with the highest total value might then be given priority for protection. (This kind of analysis could also be performed with **Geography⇒Raster⇒Combine Layers** as described in *Chapter 6*.) The **Enhance⇒Mask** operation can be used to extract an area of interest from a raster GIS layer (or from a radiometric image). For instance, you could create a mask representing the area inside the boundaries of a national park and then use this mask to eliminate non-national park areas from your other data layers.

All of the operations under **Enhance** modify the pixel values of the input images to create new images. In contrast, the *Histogram adjustment* options available in each response panel do not make any permanent changes to pixel values. Instead *Histogram adjustment* temporarily modifies the value at each pixel before sending that value to be displayed. To see the original values with no mapping, simply choose the option *None* (the default for most **Dragon** enhancement operations).

The next section provides an overview of the **Dragon** enhancement operations. Following this, you will find detailed descriptions of each operation. The chapter concludes with a set of hands-on exercises for practice doing image enhancement.

4.2 Overview

Dragon includes three types of enhancement operations: those which operate on one image to produce a new image, those which combine information from several images into a single result, and those which transform a set of image bands into a new set of images. The **Enhancement** operations are:

- Calculate the sum or average of two image bands
- Calculate the difference between two image bands
- Calculate a symmetric difference between two bands, where the resulting image values represent the absolute (unsigned) difference between the two input bands
- Calculate the ratio of one image band to another
- Calculate the normalized ratio of two image bands (a common form of vegetation index)
- Calculate the *Global Vegetation Index* of two image bands (another form of vegetation index)
- Mask one image with another: pixels in the result image are set to zero wherever the corresponding mask pixel is zero, and are otherwise unchanged
- Filter an image, with sub-options for:
  - Smooth the image to improve visibility of homogeneous regions
  - Edge enhance the image to highlight outlines and linear features
  - Median filter, to remove 'speckles' and other noise
  - User-specified 3x3 filter
- Relaxation: detect and modify pixel values that are extremely different from values of neighboring pixels
- Principle components analysis (PCA): transform a set of image bands into a new set of images which redistribute the information content of the original images. PCA is often used to create a new, smaller set of image bands for input to a classification algorithm.

Enhancement operations are most often performed on files holding raw image data. It is possible to use classified image result files as input to these operations, but normally this would not make sense. **Masking** is an exception, since the mask itself is normally a classified or coded image. Also, **Relaxation** is sometimes used to eliminate single mis-classified pixels.

All enhancement operations except PCA automatically display their results in a viewport, which the user can select. Operations that display their results provide an opportunity for the user to control contrast stretch and the color assignment (although **Dragon** will handle these aspects of the display unless the user explicitly requests a change).

With the exception of masking (where the primary objective is to apply a mask) and PCA, all enhancement operations permit using a mask file to limit processing to a specified sub-region of the image. If a mask image is specified, the enhancement operation (e.g. differencing two images) is only performed on pixels that have non-zero values in the mask.

For operations that combine two images, **Dragon** allows those images to have different dimensions. The result image will have dimensions that reflect the minimum lines and pixels per line of the two input images. Mask images can also have different dimensions than the input images. If the mask image is smaller than that input images, areas beyond the edges of the mask are processed as though there was no mask (that is, the operation is always applied to these areas). Areas of the mask which extend beyond the edges of the input image(s) are ignored (since the result image will not include these areas).

Some kinds of image enhancement require several stages of calculation to obtain a final result. In all enhancement operations, the result image data remains 'in memory' (where it can be accessed using the *special file name* =**M**), and can be combined with additional data in the next stage. It is also possible to save the results of an enhancement operation in a new image file, using **File**⇒**Save**. The saved file can be selected as input to another enhancement operation at a later stage.

The enhancement operations have many of the same common parameters as those described in the chapter discussing the **Display** operations. The major exception is that screen annotation can only be controlled within the **Display** operations.

4.3   Sum Two Image Bands (SUM)

Choose **Enhance**⇒**Sum** to add together the data values in two image bands. Usually these will represent two different spectral bands from the same scene, although as explained above, **Enhance**⇒**Sum** can also be used for raster GIS analysis, combining score data on several dimensions.

Finally, **Enhance**⇒**Sum** can be used to 'reassemble' a classified image in which different regions were classified with different methods. Suppose, for example, that you have an image that includes both land and ocean. Perhaps you want to use supervised classification (See *Chapter 5*) to classify land cover, but unsupervised classification on the water areas. One way to accomplish this would be to use the **Enhance**⇒**Mask** operation to create two sets of image bands, one that contained only land (with water pixels all set to zero) and one which contained only water (with land pixels set to zero). You could then apply classify each set of image bands using the desired method. The result would be two classified images, each of which would have zeros outside the areas of interest. By summing these two classified images, you could obtain a classified result for the entire image. Figure 4.1 illustrates this final step.

Since **Dragon** can only process image data values from 0 to 255 while addition could produce values as high as 510, **Sum** must modify its results to fit them into the range of 0 to 255. The *Scaling option* field permits you to select between two methods: either dividing the sum by 2, thus yielding an *average*, or simply truncating all values above 255 to be equal to 255. Which you choose depends on what you are trying to achieve. If you want to preserve exact data values (such as for coded files), choose *Truncate*.

You can choose to display the calculated result image either in color or in shades of gray. Different colors or gray shades will be assigned to different ranges of data values in the calculated result image. If you choose to display your results in color, you can specify what color scheme to use or let **Dragon** use a standard scheme. You can allow **Dragon** to determine an automatic contrast stretch based on image statistics, choose stretch parameters yourself, or (the default) specify no stretch in order to evaluate the raw results of the operation.

4.4   Difference of Two Image Bands (DIF)

Enhance⇒**Difference** calculates the difference between two images. The second image is subtracted from the first. Difference images are frequently used for detection of changes between one image and another.

Since **Dragon** can only process image data values from 0 to 255 while a subtraction could produce values from -255 to 255, **Difference** scales its results by dividing the calculated difference by two and then adding 128. This means that a difference of 0 between two pixels (no change) will produce an output value of 128, midway in the range of gray levels. You must keep this in mind when interpreting results.

You can choose to display the calculated result image either in color or in shades of gray. **Difference** offers the usual color and contrast stretch options.

41

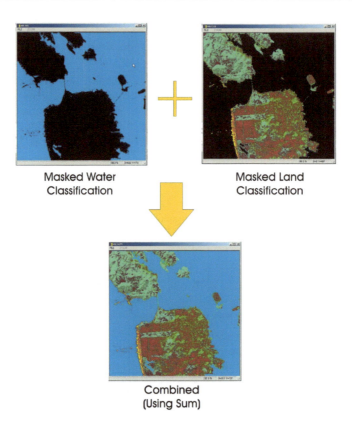

Figure 4.1: *Reassembling sub-images with SUM*

4.5 Symmetric Difference (SYM)

Choose **Enhance**⇒**Symmetric** to perform a non-directional subtraction of one image from another. The resulting image values represent how different the two images are at each pixel position, without reference to which image's value is greater. Because **Symmetric** is non-directional, it does not matter which image is selected as the first, and which as the second.

Symmetric differencing does not require any scaling, since the resulting values from input images ranging from 0 to 255 will also be in the range of 0 to 255. Thus all information about the magnitude of differences is preserved. On the other hand, information on the direction of differences is discarded. Whether you should choose a normal, directional difference or a symmetrical difference depends on your goals, on the nature of the input images, and on the type of information you are seeking. You can choose to display the calculated result image either in color or in shades of gray. **Symmetric** offers the usual color and contrast stretch options.

4.6 Ratio of Two Image Bands (RAT)

Choose **Enhance**⇒**Ratio** to divide one image band by another. This operation is directional; the first image specified is divided by the second.

Ratio operations are often used to reduce the effects of systematic distortion, such as *striping*. Striping occurs when one or more detection elements of the sensor has a higher or lower sensitivity than others. It produces a regular pattern of "stripes" across the entire image. Calculating a ratio between two bands can help this problem because the values for the stripe pixels is usually the same across all bands. Thus, stripe pixels will end up with a ratio value of 1.0.

Ratios can also reveal areas where one image band is dramatically different from another.

Ratio calculations present serious scaling problems since they can result in fractional values between 0 and 1, or in values much greater than 255. **Dragon** can only process whole number image data values from 0 to 255. The scaling method used in this operation has the following behavior: a) for any pixel where the second file (the divisor) has a value of zero, the ratio is set to 255; b) for all other pixels, the true ratio is calculated, then the result is multiplied by 16 and then converted to a whole number; c) if the result is larger than 255, it is set to 255. Thus, for those pixels where the two images are equal (the true ratio is 1.0), the result image will have a value of 16.

You can choose to display the calculated result image either in color or in shades of gray. **Ratio** offers the usual color and contrast stretch options.

4.7 Vegetation Ratio (VEG)

Choose **Enhance⇒Vegetation** to calculate the normalized ratio of one image band to another (also called the *normalized vegetation index*. **Vegetation** calculates the quantity:

```
(Band 1 - Band 2)/(Band 1 + Band 2)
```

This calculation, if performed using the appropriate image bands, is one way of computing a *vegetation index*. The brightest areas in a normalized ratio image of this type tend to be those with the highest density of vegetation. To use this operation for detecting vegetation, Band 1 should be a near IR band and band 2 should be a band from the green range of the electromagnetic spectrum.

The normalized ratio value can range from -1 to +1. To display these results, **Vegetation** scales this range to fit the image data range of 0 to 255. **Dragon** maps negative ratio values (indicating Band 1 values that are equal to or smaller than Band 2 values for a pixel) to the value 0. Positive values are multiplied by 255 to produce a value between zero and the brightest shade available.

You can display the calculated result image either in color or in gray. Because of the strongly directional nature of this calculation, gray-scale display is likely to be more satisfactory. **Vegetation** offers the usual color and contrast stretch options.

4.8 Global Vegetation Index (GVI)

Enhance⇒Global Vegetation Index (usually referred to by its abbreviation **GVI**) calculates a vegetation index according to formulae developed by the NOAA National Climatic Data Center. The GVI calculations are intended to be applied to scaled channel 1 and channel 2 data acquired from the NOAA AVHRR instrument. GVI images are particularly useful for multitemporal vegetation analyses because NOAA maintains archived GVI data stretching back more than a decade.

The GVI calculation first calculates a normalized vegetation index for each pixel according to the formula:

```
NVI = (Ch2 - Ch1)/(Ch2 + Ch1)
```

Due to previous scaling applied to the channel files, the value of this index can range from -0.05 (no green) to 0.60 (maximum green). This index is then scaled to the range 0 through 255, according to the formula:

```
GVI = 240 - (NVI - 0.05) x 350
```

This scaling inverts the index such that *the greenest areas appear darkest* in the final image. Note that this is the opposite of the **Enhance⇒Vegetation** or simple ratio methods for vegetation detection.

Enhance⇒Vegetation provides an approximation to first step of the GVI calculation. It calculates the ratio of the difference of bands to the sum of bands, then scales the result to fill the range from 0 to 255, where a vegetation ratio of 0 will be mapped to an image data value 127 in the final image.

Note that the order in which you enter data file names is different in **Vegetation** and **GVI**. **Vegetation** is intended to be used with data from a variety of different of sensors. Data values from the second image file it requests are subtracted from the first to form the numerator of the vegetation ratio. Since **GVI** is intended for use with AVHRR data commonly referred to as Channel 1 and Channel 2, **Dragon** requests the image file names in that order. Thus in the initial stage of the GVI calculation, data values from the first image file requested (Channel 1) are subtracted from the second image file (Channel 2). The help messages associated with the image file prompts each operation specify clearly the appropriate order for that operation.

4.9   Mask an Image (MAS)

Choose **Enhance⇒Mask** to highlight certain areas of a *source* image by setting other, 'irrelevant' areas to zero. A *mask* image is used to determine which pixels in the target are 'irrelevant'. Any pixel positions where the mask image has a zero value are set to zero in the result image. All other pixel positions in the result image hold the same value as in the target image. Masking is a logical rather than an arithmetic operation and thus does not require any scaling.

Mask is often used to isolate pixels belonging to particular classes. Thus, the mask image will often be a classified image, sometimes recoded (see **Classify⇒Recode**) to set irrelevant classes to 0. The target image will usually be one of the original input bands. To obtain a three-band masked image, you must mask each band separately, save each result image using **File⇒Save**, then combine the three results with **Display⇒3 Band**. You can choose to display the calculated result image either in color or gray, with the usual color and contrast stretch options. See also **Geography⇒Cookie**, which can be used to obtain a similar result based on a boundary polygon instead of a mask image.

4.10   Filter an Image (FIL)

Choose **Enhance⇒Filter** to smooth an image, edge-enhance an image, or convolve a user-specified *filter function* or *kernel* with an image. Convolution is a mathematical process in which a *filter* or *transformation* function uses information in the spatial neighborhood of a pixel to modify that pixel's value in predictable ways. Two of the most common types of filters are smoothing and edge-enhancement. **Dragon** provides these filter types automatically, but also allows you to define your own *kernels* to perform particular kinds of enhancements.

Dragon uses 3x3 filters. This means that the filtering algorithm considers the eight pixels surrounding the 'center' pixel as the spatial neighborhood:

```
    .    .    .    .    .
    .    n    n    n    .
    .    n    X    n    .
    .    n    n    n    .
    .    .    .    .    .
```

X is pixel whose value is being calculated. **n** denotes a neighborhood pixel, whose value contributes to calculations for the center pixel. The smoothing option uses a kernel of all ones, as follows:

```
    .    .    .    .    .
    .    1    1    1    .
    .    1    1    1    .
    .    1    1    1    .
    .    .    .    .    .
```

To calculate the filtered value of the 'center' pixel, each value in the filter kernel is multiplied by the value of the neighborhood pixel in the corresponding position. Similarly, the center pixel's value is multiplied by the value in the center of the filter. All these products are added together and then divided by a scaling factor, calculated as the sum of all nine values in the filter. (If the sum of filter values is zero, the scaling factor is 1.) Thus, the smoothing filter adds together 1 times the value of each of the nine pixels covered by the filter, and then divides the resulting sum by nine. In other words, it has the effect of replacing the center pixel's value with the average of that pixel's value and its eight neighbors' values.

The edge enhancement kernel is defined as follows:

```
.    .    .    .    .
.   -1   -1   -1    .
.   -1   +9   -1    .
.   -1   -1   -1    .
.    .    .    .    .
```

This filter has the effect of producing a larger value for pixels which are very different from their neighbors.

You can also define your own filter kernels. Like the smoothing and enhancement filters, these must be expressed as a set of nine positive or negative values. The specific values you use depend on what effect you are trying to accomplish. (See the *Bibliography* for sources of information on this question.) You must enter your filter values **column by column**. That is, a filter of the form:

```
.    .    .    .    .
.    1    2    3    .
.    4    5    6    .
.    7    8    9    .
.    .    .    .    .
```

should be entered in the order: 1 4 7 2 5 8 3 6 9.

As noted previously, **Filter** uses the sum of all nine filter values as a scaling factor in its computations. Even with scaling, however, the filtering calculations will sometimes result in values greater than 255 or less than 0. When this occurs, the value is set to the corresponding limit. After the filtering calculations are complete, a message indicates how many pixels were truncated (set to 0 when they were too small or 255 when too large).

In addition to the Smoothing and Edge-enhancing convolution filters, **Dragon** also provides an option for a *median filter*. Median filtering examines the same 8-pixel neighborhood around the center pixel. It replaces each pixel with the median of the values in the neighborhood. The median of a set of observed values is the value that divides the distribution in half; that is, half of all the observations have values that fall below the median, while half have values that fall above.

The effects of a median filter are similar to a smoothing filter in some ways. Pixels whose values are very different from their neighborhood are modified to be more similar to their neighbors' values. However, a smoothing filter tends to have a blurring effect on an image; sharp lines and other regions of sudden change in brightness tend to become less distinct and precise. Median filtering does a better job of preserving linear and contrasting features.

You can only filter one band at a time. To obtain a three-band filtered image, you must filter each band separately, save each of the three result images using **File⇒Save**, and then combine the three saved results in **Display⇒3 Band**.

You can choose to display the result image either in color or in shades of gray. If you choose to display your results in color, you can specify what color scheme to use or let **Dragon** use a standard scheme. You can allow **Dragon** to determine an automatic contrast stretch based on image statistics, choose stretch parameters yourself, or specify no stretch.

4.11   Relaxation (REL)

Choose **Enhance⇒Relaxation** to eliminate isolated pixels that differ sharply in value from the pixels around them. Relaxation is a process that compares each pixel in the image to its neighbors. If the neighbor pixels agree in value and the center pixel's own value is quite different from that consensus, the value of the center pixel will be changed to match its neighbors.

The general effect of relaxation is to straighten rough edges and remove high frequency noise. The results are somewhat similar to those obtained by applying a smoothing filter. However, relaxation has the advantage that it does not blur sharp edges.

45

Dragon's relaxation algorithm considers the four pixels in the same row or column as the center pixel to be the relevant spatial neighborhood:

```
        .     .     .     .     .

        .     .     n     .     .

        .     n     X     n     .

        .     .     n     .     .

        .     .     .     .     .
```

X is the pixel whose value is being calculated. **n** denotes neighborhood pixels, whose values are compared to each other and to the value of the center pixel

For each pixel in the image, the relaxation operation asks two questions: 1) Do most of the neighboring pixels have similar values; and 2) Is the center pixel similar to its neighbors? If the neighbors have similar values but the center pixel does not, the relaxation process will change the pixel to be closer to its neighbors. If either of these conditions fails, the pixel value will not be changed.

Parameters in **Relaxation** let you control the meaning of *most* and *similar* in the rules above. The *Agreement Number* parameter controls how many neighbors must agree for a change to take place. It can be set to either 3 or 4. Requiring that only three of the four neighbors agree will normally result in more pixel changes. The *Agreement Distance* parameter determines what is meant by similar. It can have a value from 0 to 7. The larger the value, the more different two values can be and still be considered similar. An Agreement Distance of 0 means that all neighbors must have identical values to be considered similar. An Agreement Distance of 7 means that neighbors' values can be up to 127 (2 to the 7th power) apart and still be considered similar!

The rule used is that if the neighbors are similar to each other, and if the center pixel is different from their average by more than the Agreement Distance, the center pixel is assumed to be in error. Its value is replaced by the average of its neighbors.

The effect of Agreement Distance on the number of pixels changed is rather difficult to predict. That is because the same definition of similarity is applied in both comparisons: the neighbors to each other and the center pixel to its neighbors. The larger the Agreement Distance, the more likely it is that the neighbor pixels will agree with each other. However, it is also more likely that the center pixel will be similar to its neighbors and thus not be changed. It is usually necessary to do some experimentation with this parameter to find the best value for a particular image. The parameter is set by default to an intermediate value.

When **Dragon** completes its calculations for relaxation, it displays the number of pixels whose values were changed. Thus you can evaluate the effects of different combinations of parameter values.

Relaxation can be repeated several times on the same image. This will frequently produce more satisfactory elimination of deviant pixels than will a single pass. To perform successive relaxations on the same data, you should choose =**M** (the default) as the image file for each pass after the first.

You can choose to display the result image either in color or in shades of gray. If you choose to display your results in color, you can specify what color scheme to use or let **Dragon** use a standard scheme. You can allow **Dragon** to determine an automatic contrast stretch based on image statistics, choose stretch parameters yourself, or specify no stretch.

4.12   Transforms

Enhance⇒Transforms provides operations to convert a set of input bands into a new set of bands, redistributing or modifying the information. In the current versions of **Dragon** this submenu includes only one operation, Principal Components Analysis (PCA).

Principal Components Analysis (PCA)

Choose **Enhance⇒Transform⇒Principal Components Analysis** to transform a set of image bands into a new set of bands (called *components*) which redistribute the information in the input images. Specifically, PCA creates new bands which have the following characteristics:

- Individual bands are independent or *orthogonal*. This means that the bands are not strongly *correlated*: the value of a pixel in one band does not predict its value in other bands. The **Utility⇒Scatterplot** operation can be used to evaluate whether image bands are orthogonal. Scatterplots of orthogonal bands will look like undirected clouds of data points. In contrast, scatterplots of correlated bands will tend to look like diagonal lines.

- By convention, the images (or components) that result from this operation are numbered from 1 up to at most (in **Dragon**'s case) 12. Lower-number components have more information, while higher-number components will tend to contain mostly noise. From a practical perspective this means that lower-numbered components, when displayed, will show a lot of detail. Higher-numbered components will look fuzzy or snowy.

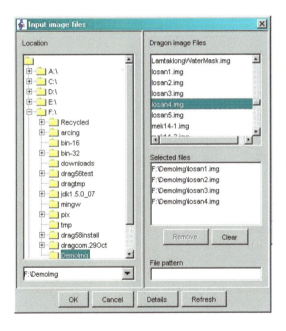

Figure 4.2: *Multiple selection file chooser*

The most common use for principal components analysis is to reduce the number of bands used as input for a classification. In a set of components derived from the seven bands of a Thematic Mapper image, the first two or three components will hold more than 90% of the information from the original seven band set. Thus, the first three components can be used for training and classification and produce results that are as good as, and sometimes better than, the original bands.

Principal components requires you to specify a set of input bands to use as input. Normally, these will be different bands or channels from the same image. **Dragon** provides a special multi-select file chooser, shown in Figure 4.2, to make the process of input band selection easier, as well as controls to allow you to change the order of input files. Figure 4.3 shows the multiple file selection field that shows up on the response panel. Each time you type a filename in the top field and press **<Enter>**, the filename will be added to the end of the list.

Dragon can process up to twelve input images.

You must also specify an output prefix, including a file path. The names for the output image files will created by adding 'C' plus the component number to this output prefix. For example, if you specify **C:\images\sanfran** as the output prefix, and you have six input images, **Dragon** will create six output images named **C:\images\sanfranC1.img**, **C:\images\sanfranC2.img**, and so on.

Figure 4.3: *Multiple selection file input field*

In addition to specifying the input and output bands, the principal components operation allows you to choose how the output components should be scaled. Scaling is necessary because the calculations involved in principal components can result in negative output values, which cannot be represented in a **Dragon** image. Scaling is also sometimes desirable in order to adjust the dynamic range of the output components. Even if the output values are positive, they will sometimes all fall within a limited range (e.g. between 0 and 32).

There are three available scaling options: **Fixed**, **Displacement** and **Multiplicative**. Fixed scaling simply adds 127 to the data values of any component with negative weightings. In this case, components with no negative values will not be changed. Displacement scaling adds a value based on the mean and standard deviation of the new data values. Values less than or equal to the mean minus three times the standard deviation are mapped to a value of zero and other values are translated accordingly. Multiplicative scaling translates values in the same way as Displacement scaling, but also scales the values so that the range of three standard deviations above and below the mean of the transformed data will be mapped to the range 0 to 255. The mean of the transformed data will be mapped to 127.

The calculations involved in principal components analysis are fairly complicated. For an explanation of these calculations, see the volume by Jensen listed in the *Bibliography*.

Principal components analysis creates output files but does not display any images. Upon completion, the operation does display a report showing the transformation matrix (eigenmatrix) used to convert the input files to the output files. You can save this report as an HTML file if you want to keep a record of these parameters.

4.13 Exercises

This section offers a set of hands-on exercises with **Dragon** enhancement operations, in order to let you practice using these functions and see their results.

1. Start your **Dragon** system. Choose **Enhance**⇒**Filter** Enter the following parameters:

   - Image File: pathum5.img
   - Kernel type: Median
   - Viewport: 0

   Click the **OK** button to start the operation. **Dragon** applies a median filter to the image (near IR band of a Landsat TM image from an area of Thailand north of Bangkok). The median filter selectively changes pixels that differ in value from the other pixels in their neighborhood.

   Choose **File**⇒**Save** and save the results of the median filtering in a file called **pathum5median.img**. Then choose **Display**⇒**Gray** and display the original file **pathum5.img** in Viewport 1. Can you see the difference? The filtered file has fewer individual bright pixels. On the other hand, you can see the linear features of canals and roads in both images.

2. Choose **Enhance**⇒**Symmetric difference**. This operation allows you to compare two images, finding areas where their pixel values differ. Enter the following parameters:

   - Image File 1: pathum5.img
   - Image File 2: pathum5median.img
   - Viewport: 0

   Click the **OK** button to start the operation. **Dragon** subtracts the second image from the first, then takes the absolute value of the difference and displays the image in shades of gray.

   The background of the resulting image is black, indicating that there was no difference between the two images in many areas. The white pixels are those whose values were changed by the median filter that you applied in step (1). Notice that these pixels are mostly along the edges of linear features like the river, as well as being concentrated in the built-up areas near the center and the lower left of the image. These are areas with *high spatial frequency information*, that is, areas where the brightness changes abruptly over short distances.

3. Choose **Enhance**⇒**Filter** again. Enter the following parameters:

- Image File: pathum5.img
- Kernel type: Edge enhancement
- Viewport: 0

Click the **OK** button to start the operation. **Dragon** applies an edge enhancement calculation to the image and displays the result in shades of gray.

Compare the result with the original **pathum5.img**, which should still be visible in Viewport 0. Notice how the details of the river's edge, the canals and the roads, that is, the edges in the scene, are all very bright. The enhancement makes some roads visible that can barely be seen in the original image, for instance, one in the upper right corner. Other details, however, are more difficult to see. The non-edge information appears a fairly uniform gray in the filtered image.

4. Choose **Enhance⇒Filter** one more time. This time we will apply a user-defined filter. Enter the following parameters, then click the **OK** button:

   - Image File: pathum5.img
   - Kernel type: User defined
   - Kernel values: -1 -1 -1 -1 8 -1 -1 -1 -1
   - Viewport: 1

The resulting image is "edge-detected" rather than just "edge-enhanced". Any non-edge pixels are set to black.

Compare the result with the edge-enhanced image which should still be visible in Viewport 0. The edge-detected (user defined filter) image shows no details at all except for edges. On the other hand, the edges are even easier to see.

5. Choose **Enhance⇒Vegetation**. Enter the following parameters then click the **OK** button:

   - Image File 1: losan4.img
   - Image File 2: losan2.img
   - Viewport: 0

The resulting image shows bright areas where vegetation occurs. Display a three band false color composite (**losan2.img** for blue, **losan3.img** for green, **losan4.img** for red), using Viewport 1. Notice how the bright areas in the vegetation index image correspond to the reddest areas in the composite.

Execute the **Enhance⇒Vegetation** operation again, using **losan5.img** as *Image File 1*, instead of **losan4.img**. Display the result in Viewport 1. What differences do you see between the two vegetation index images?

The result using band 4 distinguishes more clearly between sparse and dense vegetation. On the other hand, the result calculated with band 5 discriminates more clearly between vegetation and non-vegetation. Other than the bright areas, you can see very little detail.

6. Choose **Display⇒1 Band** and display the file **mask512.img** in Viewport 1. Notice that this file has a central area that is non-zero, surrounded by pixels with zero values. Now select **Enhance⇒Mask** Enter the following parameters then click the **OK** button:

   - Image File 1: pathum5.img
   - Image File 1: mask512.img
   - Viewport: 0

The resulting image is identical to the original **pathum5.img** in the areas where the mask image is non-zero. The result is zero where the mask is zero. Select **File⇒Save** and save the result as a file named **pathum5center.img**.

7. Choose **Display⇒1 Band** and display the file **mask512inverse.img** in Viewport 1. This file is the inverse of the previous mask. The areas that were 0 in the previous mask have a value of 255. The areas that were non-zero have been converted to zero. Choose **Enhance⇒Mask** Enter the following parameters then click the **OK** button:

   - Image File 1: pathum5.img

- Image File 1: mask512inverse.img
- Viewport: 0

The resulting image is identical to the original **pathum5.img** in the areas where the mask image is non-zero. The result is zero where the mask is zero. Select **File**⇒**Save** and save the result as a file named **pathum5surround.img**.

8. Choose **Enhance**⇒**Sum**. Enter the following parameters then click the **OK** button:

- Image File 1: pathum5center.img
- Image File 1: pathum5surround.img
- Scaling option: Truncated
- Viewport: 0

The resulting image should be identical to **pathum5.img**. We have used **Enhance**⇒**Sum** to "re-assemble" two images covering mutually exclusive areas, which were created by using masking.

9. Choose **Enhance**⇒**Difference**. Enter the following parameters then click the **OK** button:

- Image File 1: =M
- Image File 1: pathum5.img
- Histogram adjustment: None
- Viewport: 0

The resulting image is a uniform gray color, indicating that all the pixels in the original and re-assembled images are identical. (As noted above, **Enhance**⇒**Difference** scales its results, which could include negative numbers, so that -255 becomes zero and zero becomes 128.) If you used **Enhance**⇒**Symmetric difference** instead, all the pixels in the resulting image would be zero.

10. Choose **Enhance**⇒**Principal Components**, enter the following parameters then click the **OK** button:

- Input Image Files: losan1.img, losan2.img, losan3.img, losan4.img
- Output image prefix: losan
- Scaling option: Multiplicative

Dragon displays a report showing a variety of statistics about the principal components transformation. In the second section, note the line labeled **Percent variance**. Notice that the first number, corresponding to the first component, accounts for more than 75% of the variance in the data, and the second component an additional 23%. What this means is that the first two components hold almost all of the information in the original set of four image bands. The third and fourth components are mostly noise.

The four components created will be called **losan-C1.img** through **losan-C4.img** Use **Display**⇒**Gray** to display **losan-C1.img** in Viewport 0 and **losan-C2.img** in Viewport 1. Notice how different the two images are. The first component is bright in built-up areas and dark in the mountains. Roads are also bright. The second component is bright in vegetated areas. Roads are dark lines on a brighter background.

Now use **Display**⇒**Gray** to display **losan-C4.img** in Viewport 1. Very few details are visible in this image, compared to the lower numbered components. On the other hand, you can see a kind of wavy pattern of light and dark that covers the whole image. This is radiometric noise that was isolated in this component.

Chapter 5

CLASSIFICATION OPERATIONS

5.1   Principles of Image Classification

One main reason for analyzing remote sensing imagery is to gather information about conditions on the ground in the region covered by the image. The type of information that is of interest to the analyst depends on the problem she is trying to solve. For example, if the goal is to identify illegal logging in a forest reserve, the analyst will want to distinguish among undisturbed forest, forest disturbed by logging, and legitimate clearing for agriculture. If the goal is to estimate rice yields, the analyst will want to determine how much planted area is devoted to rice as opposed to other crops. To evaluate sedimentation at the mouth of a river, the analyst will want to separate clear, slightly cloudy and muddy water.

Sometimes it is possible to obtain the desired information by visual interpretation, that is, by looking at images and manually locating areas of interest. However, visual interpretation has many problems: it is labor intensive, it requires specialized knowledge, and it tends to produce different results depending on the analyst. In addition, visual interpretation often cannot make use of all the data available, especially in a multispectral image. A Landsat Enhanced Thematic Mapper image, for instance, provides eight spectral bands. However, using standard techniques, no more than three bands can be combined for visual analysis.

Automated image classification techniques are a convenient and productive approach to extracting the maximum amount of information from a multispectral image. Classification increases the effectiveness of visual interpretation. It does not necessarily replace it. The knowledge of the analyst is still critical in choosing an appropriate classification technique, identifying examples of desired classes (in supervised classification) and evaluating the results.

There are many classification techniques, but all are based on the same principle: differences on the ground generate different pixel values at the corresponding location in a digital image. Classification processes attempt to label each pixel as a member of a specific *class* based on the associated pixel value(s), often across multiple bands. As noted above, the nature of the classes will change with the problem.

Consider a very simple example. For images in the visible area of the spectrum, clear water tends to generate lower pixel values than most other categories of land cover. Thus, to identify water pixels, it may be sufficient to apply a *threshold* operation. All pixel values lower than some defined "threshold" value are labeled as "Water". All other pixels are labeled as "Land". Figure 5.1 shows an example of thresholding, which can be accomplished in **Dragon** using the **Classify⇒Recode** operation.

Usually thresholding is not accurate enough to meet the analyst's objectives. For one thing, it does not take advantage of the properties of different kinds of ground cover with respect to multiple spectral bands. If you examine the values for specific pixels across multiple bands, you will see that the pattern of *spectral response* (the amount of light reflected in different regions of the electromagnetic spectrum) varies in predictable ways. Both water and vegetation pixels will have low values in a red band. However, in a near IR band, vegetation will tend to have much higher values while water will remain low. By examining patterns of spectral response in multiple bands simultaneously, it becomes possible to distinguish between many different phenomena on the ground. Figure 5.2 illustrates this concept. (The values shown are intended to be illustrative, not accurate.)

 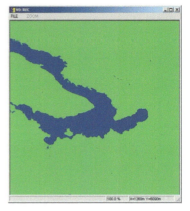

Band 4 Landsat MSS   **Threshold = 32**

Figure 5.1: *Thresholding: a simple classification method*

The distinctive pattern of spectral response for some material or ground cover class is called its *spectral signature*. Many classification techniques create mathematical descriptions of spectral signatures for each class of interest, then use some sort of rule or measure of similarity to decide, for each pixel, the class to which it belongs. These classification methods are called *supervised classification* because the analyst "supervises" or controls the process by providing pre-defined examples or specifications for each class.

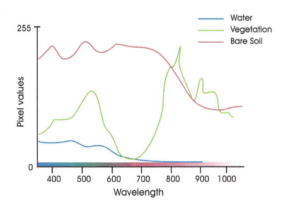

Figure 5.2: *Spectral response of different materials*

Signatures for supervised classification are usually created by selecting example points or regions on the image to be classified (or a similar/related image) and then allowing the digital image processing system to compute the mathematical descriptions. This process is called *training* the classification. The examples "train" the computer about what each class of interest "looks like". Of course, the image processing system does not actually "see" the image data. It examines the data values for each pixel to be classified and calculates the similarity between that pixel's signature and the trained signatures for each class. In priniciple, a supervised classification procedure could consider a very large number of image bands, far more than a human could ever visualize.

Some classification methods do not use training signatures. Instead, these techniques, called *unsupervised classification* techniques, identify classes by grouping pixels with similar spectral characteristics. Unsupervised classification methods discover patterns in the input data and assign pixels to classes based on those patterns. Unsupervised methods do not require any prior knowledge about the scene to be classified. This can be an advantage when it is not possible to visit the imaged region in order to independently locate training areas for each class. However, unsupervised methods cannot identify the classes they discover. The analyst must review the results and use her knowledge, plus visual interpretation, to determine what kind of material or ground cover each class represents.

The classification techniques described above are *pixel-based*. They assume that each pixel will be independently assigned to its class based only on its spectral values. In fact, the characteristics of pixels surrounding the pixel of interest can provide a great deal of information which could be used to make classification more accurate and useful. Considerable research has been devoted to the topic of *context-based classification*, that is, using the characteristics of a pixel's neighbors to influence the decision about its class. *Object-based classification* is a related topic. In the images produced by modern high resolution satellites, an object such as a building or a road may occupy multiple pixels. Pixel-based classification is of little use for such detailed imagery. Object-based classification uses shape, texture and other higher-level characteristics to group pixels into real-world objects.

Dragon does not provide any context-based or object-based classification methods. However the **Dragon** Programmer's Toolkit can be used to explore these more sophisticated approaches.

5.2 Overview of Dragon Classification

Classify includes operations to set up and perform supervised and unsupervised classification on a set of image bands. *Supervised* classification methods categorize each pixel of a scene into one of several mutually exclusive classes, based on comparisons with examples of each class that you must provide. *Unsupervised* classification methods also categorize pixels into several classes, but in contrast to supervised methods, the unsupervised methods create the class definitions automatically.

Many of the classification operations are used to define, examine and modify these examples, which are usually called *training sets* or *training signatures* (because they 'train' the classification process as to what each class 'looks like'). After creating the training signatures, you use a supervised classification procedure to automatically perform the assignment of pixels to classes.

Dragon provides three methods for supervised classification: *boxcar* or *parallelepiped*, *minimum distance to mean* (MDM) and *maximum likelihood*. These methods use the same training signatures, but they decide on class membership using different calculations and decision rules.

Dragon also provides a type of unsupervised classification procedure called *clustering*. No training signatures are required for clustering. A first guess regarding typical signatures for each of several class is refined through repeated classification attempts. Two additional operations are included under the category of unsupervised classification: a density slicing operation to provide a simple division of the image values into a small number of groups, and the procedure known as *agroclimatic assessment* (*AGC* for short).

The output of both supervised and unsupervised classification processes is an image coded by classes. This image is displayed when the classification is complete; the results can be saved in a classified image file with **File⇒Save**. **Classify** also includes the **Recode** operation for subsequent combining or renumbering of classes and the **Error Analysis** operation for evaluating the accuracy of classification results.

The **Classify** operations include:

- Define training signatures by:
    - Identifying areas on an image that represent each class
    - Applying previously-identified area boundaries to a new image
    - Using another image, called a *class map*, to define training pixel locations

- List the training signatures stored in a file

- Edit the training signatures by:
    - Displaying signature contents (signature statistics)
    - Copying a signature from another file
    - Renaming a signature
    - Deleting a signature
    - Combining two signatures
    - Modifying signature statistics
    - Displaying histograms of each band of a signature
    - Saving signatures in a file

- Supervised classification:
    - Boxcar classification
    - Minimum distance to mean
    - Maximum likelihood

- Unsupervised classification:
  - Clustering
  - "Slicing" an image into equal-sized classes
  - Agroclimatic assessment.

- Recoding a saved classified image to change class numbers, combine classes, etc.

- Analyzing the errors in a saved classified image by comparing the results to a set of standard values assumed to be correct.

Classification is one of the most complex areas in digital image processing. Successful classification depends much more on your knowledge and judgment than does successful enhancement. We have tried to make **Dragon**'s classification operations as easy to use as possible; however, you still need a basic understanding of classification concepts in order to use these operations correctly.

Most classification operations can be constrained by a mask image. This can be very useful, since it allows you to classify different regions of an image using different sets of signatures or different classification algorithms, and then combine the results (using **Enhance**⇒**Sum**, for example). It also permits you to restrict classification computations to a study area, which can result in significant time savings in time-consuming operations such as maximum likelihood classification.

5.3 CLASSIFIED IMAGE CHARACTERISTICS

The results of the classification process are normally saved in a *classified image file*. Classified image files have essentially the same structure as any other image file. However, they do have a number of special characteristics and are processed slightly differently from normal (radiometric) image files in certain circumstances. Classified image files are discussed in *Chapter 2*; their special characteristics and processing are summarized briefly below:

- Values in a classified file should be viewed as identifying labels, not measurements. Usually it does not make sense to perform arithmetic operations on classified image files.

- Classified image files include extra class-name fields in their file header. These fields can be displayed or modified using operations in the *Utility Operations* chapter.

- Classified images are never subjected to any kind of contrast stretch in Display operations.

- Each classified image file has an associated *color scheme*. Initially, this color scheme reflects the system default colors, but it may be redefined using **Utility**⇒**Colors**. Classified images are always displayed using their associated color schemes, unless you explicitly request a different color scheme. You can change the color scheme associated with classified images, or even associate a color scheme with a non-classified image, using **Utility**⇒**Header**.

5.4 Training Signature Creation (TRA)

Choose one of the **Classify**⇒**Training** operations to create a new set of training signatures for use in classification. There are three methods for doing this:

- Define training areas on an image
- Apply previously-defined training area boundaries to a new scene
- Automatic signature generation

Define Training Areas (DEF)

Choose **Classify**⇒**Training**⇒**Define** to interactively define sample areas that belong to each class, read their data values from the appropriate image files, and calculate signature statistics for each class. You can save the training area boundaries in a polygon file, if you wish, so they can be used later with **Apply**. You can also include polygons defined during a previous training session.

Define leaves the calculated training signature statistics in memory, where they can be used right away by choosing **Boxcar**, **Maximum Likelihood**, or **Minimum Distance to Mean**. They can also be viewed, modified and saved (in a signature file) using the **Classify⇒Edit Signatures** operations.

The supervised classification methods provided by **Dragon** are multispectral; patterns of values across a set of spectral bands are used to recognize particular classes. Usually the larger the number of independent bands used in a classification, the better the results. (You can use **Utility⇒Scatterplot** to determine whether two bands are independent.)

You must use at least two bands for your training signatures. Note that you must use the same number of bands (and the same sequence of bands) in training set definition as in your final classification. Thus you should evaluate the independence of different spectral bands before you begin selecting training areas.

You must select an image file for each band you want to use. It is important that you remember or record the order in which you name the files in **Define**; you must name them in exactly the same order when you actually perform the classification or your results will be erroneous.

Define permits you to select the image you will view while defining training areas. **Dragon** asks you for an image file to serve as a *background image*. This background image is the image you will see on the screen while you are selecting the training area locations. The current image in memory, if any, is the default. You can also enter =**C** to use the last-displayed three band composite (if any) as the background.

After you have provided values for all parameters, the interactive training process begins. The selected image is displayed in the training viewport.

In **OpenDragon**, you can begin choosing training areas immediately. However, in **Dragon Professional** the image will initially be displayed as an *overview*. You must first choose a *region of interest*, as described in *Chapter 2*. The region of interest will then be displayed at a 100% (1-to-1) resolution. You may return to the overview (using the right mouse button) and select a new region of interest as often as needed during the training process.

During the training process, **Dragon** also displays a button panel that allows you to select various operations. The operations on the button panel can also be invoked via function keys. The button panel provides the following options:

- **Accept** (<**F1**>): completes the training interaction and continues with signature calculations.
- **Help** (<**F4**>): displays general instructions for training.
- **New Class** (<**F6**>): Switches to defining training areas for a different class. This can be used to define a new class, or to return to a previously defined class. The option brings up a dialog which prompts for a class name. This dialog also contains a list from which you can select a previously used class.
- **Delete** (<**F9**>): Deletes the last created training area.
- **Cancel** (<**F10**>): Exits from the training process without computing signature statistics or saving polygons.

Buttons that are not relevant are disabled. For example, the **Delete** button will not be enabled until you define a training area.

To start the training process, click on a point which will be the center for a training area. **Dragon** will display the Class Name dialog, allowing you to enter the name of your first class. Class names may be up to thirty-two characters long, and can include letters, numbers, dashes, underscores and spaces - in fact, any printable Unicode character. (You can also begin by choosing <**F6**> or clicking on the **New Class**.)

Once you have entered a name for the first class, and clicked on **Ok**, you are ready to start defining training areas. Training areas created using **Define** are circular. Thus, to define a training area, choose two points: a center for the circle, and a second point whose distance from the first defines the circle radius. Messages at the bottom of the viewport tell you what to do at the appropriate times.

To select center or radius points, position the mouse cursor in the desired location, using the mouse, and the click the left mouse button, or press the <**Ins**> key. Fields at the bottom of the viewport display continuously updated X (pixel) and Y (line) coordinates of the cursor. Note that when the mouse cursor is inside the viewport, you can also move mouse cursor one pixel at a time using the arrow keys on the keypad.

When you press the <Ins> key or the left mouse button to select a circle center, a little X will appear at the point you have chosen. When you press period again, to define the radius, the center point disappears and the circle is drawn. All pixels within that circle or on its edge will be used to calculate the training signature for the class you chose previously.

If the circle encloses only pixels you want to include in your training set, you can go on and choose a new center and radius. If the circle encloses pixels that should not be included in the training set, you should press click on the **Delete** button or press <F9>, to make the circle disappear. In either case, you can continue to define circles, all of which will be used together to calculate a training signature for the most recently-selected class.

To select points more precisely during training, you may *zoom* (enlarge) the image on the screen. To zoom the image, choose one of the options from the viewport **Zoom** menu. Then move the cursor into the general area where you want to choose a training area and click the left mouse button. The image will be redisplayed, enlarged. The place clicked will become the center of the zoomed image.

In a zoomed image, it is easy to see individual pixels. Thus you can locate the cursor very precisely. You can also see very easily when pixels of the wrong color have been included in a circle.

When you want to return to the normal, non-zoomed image, choose **Zoom⇒Restore**. The image will be redisplayed at its original size. All previously-defined circles for the current class will be redisplayed.

To switch to defining training areas for a different class, press <F6> or click on the **New Class** button. **Dragon** will prompt you to enter a new class name; alternatively, you can select a previously defined class from the list displayed in the dialog.

When you have completed training area definition for all your classes, press the <F1> key or click on the **Accept** button. **Dragon** will read the image files that you have specified, calculating statistics for each class across all the files. The signature statistics will be available immediately upon completion of the training operation. You can examine or modify them using the **Classify⇒Edit Signatures** operations, by choosing (**Current**) as the name of the signature file. You can also use them in any of the supervised classification operations.

When the signature calculations are complete, **Dragon** displays a new image. This image shows all the training areas that you defined, as colored circles on a black background. Each class is shown in a different color. You can save this image, using **File⇒Save**, and then use it as a class map file in **Classify⇒Training⇒Automatic**.

In addition to calculating and retaining signature statistics, **Dragon** also saves boundary information for all the circles you define in a training session. You can specify a filename for saving these boundaries by entering a valid filename for the *Polygon Save File* parameter. Even if you do not specify a filename, **Dragon** saves this information in a temporary file. To use these boundaries immediately to continue an interactive training session, or as input to **Apply**, specify (**Current**) for the *Polygon File to apply* parameter.

If for some reason you decide to abandon your training session without calculating signatures, press <F10> or click on the **Cancel** button. Under these circumstances, the training area boundaries will not be saved.

Apply Training Area Boundaries (APP)

Choose **Classify⇒Training⇒Apply** if you previously saved training area boundaries in **Define** and you now want to use those boundaries to calculate training signature statistics. **Apply** makes it possible to define identical training area locations on several different images. One common use of stored training boundaries is to calculate several sets of training signatures, using different groupings of bands but the same training area locations. The saved training area boundaries allow you to compare classifications of the same scene using different groupings of bands, without repeating the interactive training process for each classification.

Usually the larger the number of independent bands used in a classification, the better the results. (You can use **Utility⇒Scatterplot** to determine whether two bands are independent.) As in **Define**, you must specify at least two bands, and you must use the same number of bands during training set definition as in your final classification.

You specify one image for each band requested. **Dragon** calculates the total number of bands based on the number of images you supply. Normally these image bands should all represent the same scene. It is important that you remember or record the order in which you name the files in **Apply**. You must name them in exactly the same order when you actually perform the classification or your results will be erroneous.

Finally, **Apply** requires the name of the file where the training area boundaries were saved.

Once you have provided values for all parameters, **Dragon** reads the previously-defined circle information from the polygon file. Then the program reads the image data for each training area from each image file and calculates signature statistics for each class.

As in **Define**, when the signature calculations are complete, **Dragon** displays an image showing all the training areas defined in the input polygon file, as colored circles on a black background.

Automatic Signature Generation (AUT)

Choose **Classify⇒Training⇒Automatic Signature Generation** (**AUT**) to extract signatures to be used for supervised classification from a set of image bands. Rather than using circular training areas to specify which pixels belong in each signature, **Automatic Signature Generation** uses a *class map* image to define the location of training points. The class map can be any image, from any source. In particular, the class map can be one of the following:

- A classified image generated by the clustering operation. This allows supervised classification signatures to be developed based on unsupervised classification results.
- A layer file created by defining and filling polygons in **Geography⇒Vector**. This provides a method for defining non-circular training areas.
- A thematic image such as a soils map imported from a geographic information system (GIS). This provides a direct method for integrating ancillary information into the classification training process.
- The image that results from running **Define** or **Apply**, which represents training areas as filled circles. This provides another method of applying a set of trained signatures to a different set of image data bands.

Note that the class map image can be subjected to various operations including masking, recoding and editing via **Geography⇒Vector**, before it is used. **Automatic Signature Generation** thus gives users much more control over the content of training signatures than is available in the **Define** and **Apply** operations.

Signature	Class Map Value
1	60
2	80
3	100
4	120

Figure 5.3: *Equivalences between class map values and signatures*

Automatic Signature Generation reads up to four sample data files, collecting signature statistics for pixel locations that have non-zero values in the class map image. Each distinct value in the class map image defines a separate class. Signatures can be generated for up to 16 classes. If the class map image includes more than 16 distinct values, the 16 lowest values are used to define the location of training points; other values are ignored (treated as if they were zeros).

When **Automatic Signature Generation** completes the process of reading the input image files and calculating the signature statistics, it displays a simple table indicating which values in the class map image were used to define each signature. An example is shown in Figure 5.3. This table is particularly useful when the class map image is a layer file or a GIS layer with a wide range of values, rather than derived from a previously-classified image.

The signatures produced by **Automatic Signature Generation** are essentially identical to those generated through the **Define** and **Apply** operations. They can be used in all the **Edit Signatures** operations, saved in signature files, and used in all supervised classification operations.

Note, however, that depending on the structure of the class map image, automatically-generated signatures may represent a much larger number of points than explicitly-trained signatures. For instance, a signature derived from clustering results may represent tens or even hundreds of thousands of pixels. Normally, this should cause no problems. **Classify**⇒**Edit Signatures**⇒**Combine** will give an error message if you try to combine two signatures that represent widely-differing numbers of pixels, since in this case the weighting methods used in **Combine** are not statistically valid.

The *SOURCE* field in the signature file, which can have the value *TRAINED* or *AUTO*, provides a method for determining the origin of a particular signature. This field is set automatically during signature creation.

5.5   List Signatures (LIS)

Choose **Classify**⇒**List** to view the names of signatures stored in a signature file, or for the latest calculated or used set of signatures (the 'current' signatures). This operation can help you find a particular signature file that you want when you have several files available.

5.6   Edit Signatures

Choose one of the **Classify**⇒**Edit Signatures** operations to display and modify information contained in training signatures. **Edit Signatures** offers a variety of signature review and manipulation operations including:

- Display signature contents (signature statistics)
- Copy a signature from the secondary file
- Rename a signature
- Delete a signature
- Combine two signatures
- Modify signature statistics
- Display histograms of a signature
- Save signatures in a file

All operations except for **Save** require you to specify a signature file (referred to as *Primary Signature File* in the cases of **Copy** and **Combine**). This can be **(Current)** if there is already a set of signatures in memory. Whatever you specify will become the currently in-memory signatures. Up to sixteen signatures (representing sixteen classes) can be in memory at any time.

Most **Edit Signatures** functions operate only on the signatures in memory. **Combine**, however, can add a signature from a secondary file to one in memory. It can also add one signature in memory to another. **Copy** can move a signature from a secondary file into memory, if there is room for it.

The results of signature editing operations exist only in memory unless they are explicitly saved in a signature file. After a **Classify**⇒**Edit Signatures**⇒**Save** operation, the signatures are still available in memory, so that further modifications can be made and saved in a signature file of a different name. In fact, signatures remain in memory and can be further modified even after classification has been performed.

Most **Edit Signatures** operations require signature names as parameter values. Signature names must be typed exactly as they appear in the list of available signatures on screen during the **Edit Signatures** interactions. Alternatively, you can choose existing signatures from the drop-down lists on the screens for each operation. This list will be populated with the appropriate signature names for a particular signature file, as soon as you tab off the signature file field.

Show Signature Statistics (SHO)

Choose **Classify**⇒**Edit Signatures**⇒**Show** to display in tabular form the contents of signatures in memory or in a signature file. The statistics listing shows the signature name, the number of pixels contributing to the calculated values, and then, for each band, the mean, variance, maximum data value, minimum data value, upper bound and lower bound, and the name of the image file from which the data were gathered. It also displays the covariance of each pair of bands, in matrix form. Figure 5.4 shows an example of the output produced by **Classify⇒Edit Signatures⇒Show**.

The covariance of two bands measures the degree of correspondence or dependence between the bands. Large covariance values (significantly larger than 1.00, positive or negative) indicate that the two bands are not independent, but carry redundant information. If two bands have a very large covariance, you may want to eliminate one of them from your classification.

The upper and lower bounds are used as decision limits in assigning pixels during boxcar classification. The boxcar method assigns a pixel to a class if and only if the pixel's value in each band lies between the upper and lower bounds for the band, for that class. When signature statistics are first calculated, the upper and lower bounds are set equal to the maximum and minimum values in the training set, respectively. However, the upper and lower bounds can be changed using **Classify⇒Edit Signatures⇒Modify**.

The mean, variance, maximum and minimum values are not used directly in boxcar classification. However, you can use these statistics to compare signatures and estimate the degree of separability between the classes you have defined. Ideally, signatures intended to represent different classes should have distinctly different means and non-overlapping ranges in at least one band.

You can also use the mean and variance to calculate reasonable upper and lower bounds for use in **Modify**. According to the rules of probability, 98% of all values in a sample should fall within two standard deviations on either side of the mean. Thus, a reasonable upper bound would be the mean value plus two times the standard deviation; a good lower bound might be the mean minus twice the standard deviation. If the variance of a signature is very large, this rule of thumb will give unreasonable results (e.g. the upper bound at 255 and the lower bound at 0). However, a large variance frequently indicates that your sample includes more than one spectral class and should probably be discarded.

The minimum distance to mean classification method uses the mean values for each signature to characterize the class. This method assigns each pixel to the class whose means for each band are closest to that pixel's data values on each band. The other calculated signature statistics are not used in the minimum distance to mean procedure.

The maximum likelihood classification method uses the signature means, variances, and covariances in its calculations. Maximum likelihood assumes that the data values for each class approximate a standard normal distribution (Gaussian distribution). Under this assumption, the method calculates the probability that a pixel belongs to each class, and assigns the pixel to the most likely class.

Show lets you examine a specific signature or all the signatures in memory. To view all signatures, select "*" from the Signature Names drop-down list.

Modify a Signature (MOD)

Choose **Classify⇒Edit Signatures⇒Modify** to change the upper and/or lower bounds for one or more bands in a signature. Any bounds not changed will remain at their previous values.

The upper and lower bounds are used as decision limits in assigning pixels during boxcar classification. The boxcar method assigns a pixel to a class if and only if the pixel's value in each band lies between the upper and lower bounds for the band in the signature for that class. When signature statistics are first calculated, the upper and lower bounds are set equal to the maximum and minimum values in the training set, respectively. However, the maximum or minimum values may be outliers (that is, extreme values that do not accurately represent the distribution of values in the signature). Outliers can be detected by viewing the histograms of signature bands. More representative values for the bounds can be determined using the statistics displayed by **Show**.

Changing the upper and lower bounds will have no effect on minimum distance to mean or maximum likelihood classification. There is no way to modify the signature statistics directly to adjust the results of these methods. If your signatures appear to be heavily biased by outliers, you should probably discard them and define a new set of training areas.

Signature Name Harbor			
Number of Points 3235		**Source** TRAINED	

	Image boston1	**Mean** 48	**Variance** 2.786000
Band 1	**Min** 44 **Max** 52	**Lower Bound** 44	**Upper Bound** 52

	Image boston2	**Mean** 31	**Variance** 3.320000
Band 2	**Min** 26 **Max** 35	**Lower Bound** 26	**Upper Bound** 35

	Image boston4	**Mean** 13	**Variance** 0.293000
Band 3	**Min** 11 **Max** 14	**Lower Bound** 11	**Upper Bound** 14

	Image boston5	**Mean** 10	**Variance** 0.442000
Band 4	**Min** 7 **Max** 12	**Lower Bound** 7	**Upper Bound** 12

Covariance matrix:

2.786	2.168	0.137	0.049
2.168	3.320	0.200	0.051
0.137	0.200	0.293	0.001
0.049	0.051	0.001	0.442

Figure 5.4: *Typical output from Show Signatures*

Copy a Signature (COP)

Choose **Classify**⇒**Edit Signatures**⇒**Copy** to transfer a signature from a secondary signature file to memory. You can execute **Copy** only if there are fewer than sixteen signatures already in memory.

Copy is useful for combining signatures from two different training sessions, representing two sets of classes, into a single signature file.

Combine Two Signatures (COM)

Choose **Classify**⇒**Edit Signatures**⇒**Combine** to add together two signatures. Normally, the two signatures should represent the same class in order for the results to be meaningful, but **Dragon** cannot enforce this. **Combine** allows you to combine a signature from the secondary file previously specified with a signature in memory. Thus, you can combine signatures representing the same class taken from two different images, in two separate training sessions.

To use **Combine**, specify three signature names: the name of an initial signature, which must reside in memory; the name of a signature to combine with the first, which can be either in memory or in the secondary file; and a name for the resulting combined signature. The combined signature will take the place of the first signature specified. The second signature will not be changed. However, if it also resides in memory, you probably should eliminate it using **Delete**, since it carries redundant information already included in the combined signature.

Dragon combines signature statistics according to the following rules. The number of pixels in the combined signature is the sum of the number of pixels in each contributing signature. For maximum, minimum, upper bound and lower bound, **Dragon** compares the corresponding values for the two signatures, and uses the one that is the more extreme (larger for maximum and upper bound, smaller for minimum and lower bound). **Dragon** combines the histograms simply by adding the frequencies for each value. Finally, **Dragon** uses a weighted averaging to combine the means, variances, and covariances. The weights used are proportional to the number of pixels in each signature.

Rename a Signature (REN)

Choose **Classify**⇒**Edit Signatures**⇒**Rename** to change the name of a signature in memory. Signature names should provide information on the class the signature is intended to represent. Signature names can be up to thirty-two characters long and can contain letters, numbers, dashes, underscores and blanks.

Rename requires you to provide old and new signature names. The new signature name will replace the old name.

Note

Avoid giving the same signature name to two different signatures in memory. If duplicate signature names exist, only the first signature can be accessed.

Save Signatures in a File (SAV)

Choose **Classify**⇒**Edit Signatures**⇒**Save** to store all signatures currently in memory in a signature (**.sig**) file. This signature file can be used later for classification or as input to the Edit Signatures operation.

The one parameter you must enter is the name of the signature file to be created. **Save** does not change the signatures in memory. Thus you can make some changes, save a version of the signatures, make some more changes, save a new version in a file with a different name, etc.

Delete a Signature (DEL)

Choose **Classify**⇒**Edit Signatures**⇒**Delete** to eliminate a signature from memory. **Delete** is most often used to make room for new signatures to be copied from the secondary file. **Delete** is also used frequently to eliminate a signature after it has been combined with another signature, to remove redundant information, or to get rid of a signature which has an unacceptably high variance or other statistical problems.

Histograms of Signatures (HIS)

Choose **Classify**⇒**Edit Signatures**⇒**Histogram** to display in graphic form the frequency distribution of data values in all bands of a signature. Such frequency graphs are called *histograms*. The horizontal axis in the histogram represents the different possible data values. The vertical axis in the histogram represents the number of pixels in the signature that have each data value. Figure 5.5 shows typical signature histograms produced by **Dragon**.

Histograms of training signatures are valuable tools for a variety of purposes. The histogram can help you identify the presence of outliers (extreme and probably spurious data values not representative of the signature as a whole). You can also use histograms of training signatures to detect bimodal distributions of values, that is, distributions in which there are two peaks of high frequency rather than one. A bimodal distribution usually indicates that the signature contains pixels from two classes and should be discarded.

The width of the peaks in signature histograms provides a graphic indication of the consistency of values in the signature. You may want to discard signatures with either extremely narrow and extremely broad peaks. A signature with a very narrow peak may not represent the full variability of the class and may result in many unclassified pixels. A signature with a very broad peak may actually be a mixture of several classes. (For more information on the interpretation of signature histograms, see references in the *Bibliography*.)

To display signature histograms, you must provide the name of the signature you wish to display. Histograms for all signature bands are displayed simultaneously.

5.7   Supervised Classification (SUP)

Classify⇒**Supervised** provides operations to classify a scene into categories defined by training signatures.

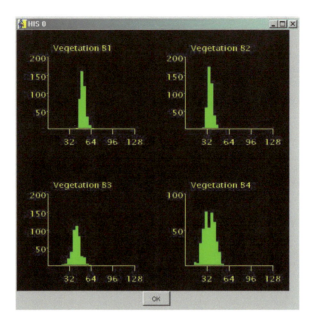

Figure 5.5: *Example of signature histograms*

Boxcar Classification (BOX)

Choose **Classify⇒Supervised⇒Boxcar** to classify a scene into categories defined by training signatures using the boxcar method and to display the classified image.

The boxcar (or parallelepiped) method uses the upper and lower bounds of each band in a signature to define class membership. If a pixel's values in each band fall within the range defined by the upper and lower bounds for that band, the method assigns the pixel to the class represented by that signature. Pixels whose values fall outside the ranges of all signatures on one or more bands are not assigned to any class; they are called *unclassified*.

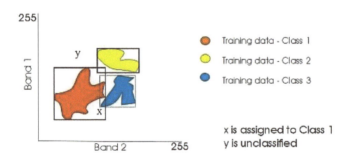

Figure 5.6: *Boxcar classification*

Figure 5.6 shows a simplified example where there are only two bands of input data. The black rectangles around each blob illustrate the upper and lower bounds of each set of training data. The sample data point at X would be classified as class 1 because it is inside the rectangle for that class (and not any other classes). Since the sample data point at Y does not lie within the boundaries of any training data set, it is unclassified.

If there is no overlap among signature ranges, **Boxcar** provides very clear results. However, if signature ranges for different classes do overlap, then class assignment can be ambiguous. A pixel whose values fall in a region of overlap could belong to several classes. Such pixels are known as mixed pixels. A mixed pixel does not usually represent a mixture of classes on the ground. However, it is not possible to determine, without additional information, the 'true' class of mixed pixels. Therefore, **Boxcar** simply identifies these pixels by their possible classes and lets you decide how to handle them. (Frequently ancillary information from maps or ground surveys can help you clarify the true class of mixed pixels.)

In Figure 5.6, there are only small areas of overlap between any pair of signatures. This means that this classification would produce few mixed pixels.

To perform a **Boxcar** classification, you must have a set of training signatures, either in a signature file or in memory. The signatures will be in memory if you have just finished executing a **Training** or **Edit Signatures** operations.

Boxcar classification is a type of multispectral classification: patterns of values across a set of spectral bands are used to recognize particular classes. Usually the larger the number of independent bands used in a classification, the better the results. (You can use **Utility**⇒**Scatterplot** to determine whether two bands are independent.) You must specify at least two bands. You should use the same number of bands in your final classification as you used in defining your training signatures.

You must specify an image for each band requested. **Dragon** calculates the number of required bands based on the signature file you have selected. Normally these image bands should all represent the same scene. If you are using the same image bands that you used to define your training signatures, it is important that you name them in the same order as when you created the signatures, or your results will be erroneous. If you used a different scene for developing your training samples, you should still name the image files according to the same ordering of bands. That is, if you developed your training sets on Scene A, using Thematic Mapper Bands 2, 3 and 5, and you now want to classify Scene B, you should use the Scene B files representing Bands 2, 3 and 5, in that order.

Once you have provided values for all parameters, **Dragon** reads the signatures from the signature file (if necessary) and then reads each image file in turn, proceeding with the classification. When the classification is complete, **Dragon** displays a table showing the number of pixels and percentage of area in each pure class and the number of mixed pixels in each confusion group (set of classes that have some overlap). Confusion groups are identified in the table with expressions such as: 1+3+4. This would mean that pixels in that group might belong to either class 1, class 3 or class 4; you can't tell which without additional information. A sample table is shown in Figure 5.7.

Class/Cluster	# Pixels	Area (%)	Composition
Unclassified	70467	7.04	0
1 Water	228	0.02	1
2 Vegetation	19953	1.99	2
3 BareRock	44229	4.42	3
4 BareSoil	2231	0.22	4
5 Scrub	35733	3.57	5
6 Concrete	24425	2.44	6
7	179065	17.90	2+4+5
8	176922	17.69	2+3+5
9	129742	12.97	4+5
10	109389	10.93	4+5+6
11	75695	7.56	2+5
12	54440	5.44	5+6
13	27505	2.75	2+3+4+5
14	23733	2.37	2+3
Total	1000000		

Figure 5.7: *Class frequency table for boxcar classifier*

Note that although other classification operations permit you to use up to 16 classes, **Boxcar** is limited to only eight pure classes. If your signature set includes more than eight signatures, **Boxcar** will use only the first eight.

When **Boxcar** reports the results of its classification, it will report the unclassified area as class 0, then the areas for up to eight pure classes, then the areas for the seven highest frequency mixed-pixel classes, and finally the total area for all remaining mixed-pixel classes (none, in figure 5.7).

After the frequency table appears, the classified image is displayed. A standard color scheme is used.

Note that even after classification is complete, the signatures used remain available in memory. Thus, if you are dissatisfied with your first classification results, you can go back to **Classify**⇒**Edit Signatures** and modify the bounds, then try the classification again. You can also save your classification results with **Utility**⇒**Save**, and then run a new classification using the same signatures but different image files.

Maximum Likelihood (MAX)

Choose **Classify**⇒**Supervised**⇒**Maximum Likelihood** to classify a scene into categories defined by training signatures using the maximum likelihood method, and to display the classified image in a selected viewport.

Maximum likelihood classification uses computed probabilities to define class membership. It calculates the probability that a pixel belongs to each class, using the formula for probabilities drawn from a multivariate normal distribution. The method then assigns the pixel to the class with the highest probability. To calculate the probabilities, **Maximum Likelihood** uses the signature statistics: means, variances and covariances. Unlike **Boxcar**, **Maximum Likelihood** by default assigns every pixel to some class; it does not produce unclassified pixels. However, using the *Threshold* parameter, you can specify that the most deviant pixels should be left unclassified.

Unlike the other classification methods provided, **Maximum Likelihood** makes some assumptions about your signatures. Specifically, it assumes that the individual training point data values in each signature and band are distributed according to a normal (Gaussian) distribution. If this assumption is seriously violated, maximum likelihood will give erroneous results. You can evaluate the appropriateness of the normality assumption by examining the histograms for each signature. If the signature histograms are unimodal (have only one peak) and symmetrical (roughly the same size and shape on either side of the peak), then maximum likelihood classification can be undertaken. If some of the histograms are seriously skewed (have more points on one side of the peak than the other), you should use another classification method, or repeat the training process to obtain signatures that more closely fit the normality assumption. Examples of approximately normal, bimodal and high variance, somewhat skewed signature histograms are shown in Figure 5.8.

A comprehensive discussion of maximum likelihood classification is beyond the scope of this manual. The *Bibliography* lists several texts that can provide more complete information. There are, however, a few important points to make in the context of using maximum likelihood in **Dragon**.

Compared to other **Dragon** operations, maximum likelihood classification requires a complicated set of calculations to be carried out for each pixel. Thus, it is much slower than most other operations. Weighed against this disadvantage is a powerful advantage: maximum likelihood classification is an optimal classification method. If its assumptions are met, maximum likelihood produces a classification solution that minimizes classification errors.

Maximum likelihood classification is most useful when signatures for different classes overlap in spectral space. When class assignment is ambiguous, maximum likelihood identifies the most likely class for pixels whose values fall in a region of overlap. Thus, it provides an optimal class assignment for pixels that would be identified as "mixed" in the boxcar method.

If the classes of interest do not overlap significantly, there is little advantage in using maximum likelihood classification. One of the other methods will probably give comparable results, in a considerably shorter time.

Figure 5.9 shows the same simplified example where there are only two bands of input data. The black dots represent the signature means and the black ovals are intended to indicate the variance within each signature.

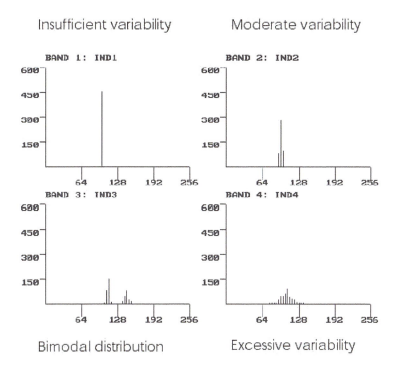

Ideally, signature histograms should have a single peak, be roughly symmetric, and show moderate variability in values.

Figure 5.8: *Evaluating signatures based on their histograms*

Note that the calculation of probabilities used in Maximum Likelihood considers all the information available in a signature: its average value in each band (mean), the variance within each band (degree of variability) and the degree to which the variation in each band is correlated with other bands (covariance, not illustrated in the figure). The diagram shows that even though point y is slightly closer to the mean of class 2, it has a higher probability of belonging to class 1 because the class 1 signature has a high variability.

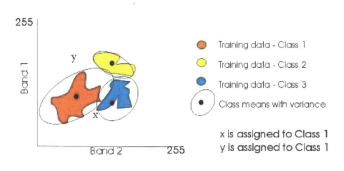

One of the computations required by maximum likelihood is inversion of the covariance matrix for each class. Occasionally, one or more of the covariance matrices will be *singular*, that is, impossible to invert. When this occurs, classification cannot proceed. **Dragon** gives you an error message and aborts the process. Usually, this problem can be remedied by adding more training points to the signature that displays this problem, or by discarding the signature altogether and gathering new training data for that class. See the sections on **Classify⇒Training** and

Figure 5.9: *Maximum likelihood classification*

Classify⇒Edit Signatures for further information.

To perform a maximum likelihood classification, you must have a set of training signatures, either in a signature file or in memory. The signatures will be in memory if you have just finished executing **Classify⇒Training** or **Classify⇒Edit Signatures**.

Maximum Likelihood requires you to select an image for each band you want to classify. **Dragon** computes the number of required bands based on the signature file you have selected. It is important that your image files represent the same bands as you used in training, in the same order as when you created the signatures.

Maximum Likelihood allows you to specify a threshold as a real number between 1.0 and 10.00. This value is used to calculate per-class thresholds, based on the standard deviations of the pixels in the class. The larger the number, the more lenient the threshold, and the fewer pixels that will be excluded as unclassified.

Once you have provided values for all parameters, **Dragon** reads the signatures from the signature file (if necessary) and then proceeds with the classification. When the classification is complete, **Dragon** displays a table showing the number of pixels in each class. Then the classified image is displayed. A standard color scheme is used.

Note that even after classification is complete, the signatures used remain available in memory. Thus, you can save your classification results and then run a new classification using the same signatures but different image files.

Minimum Distance to Mean (MDM)

Choose **Classify⇒Supervised⇒Minimum Distance to Mean** (usually referred to by its abbreviation **MDM**) to classify a scene into categories defined by training signatures using the minimum distance to mean method and to display the classified image in a selected viewport.

MDM uses relative distance from the signature means to define class membership. **Dragon** uses Euclidean distance to measure how close a pixel is to each signature mean. For each pixel in an image, **Dragon** computes the distance in spectral space from each set of signature means. Then it assigns the pixel to the class corresponding to the closest mean.

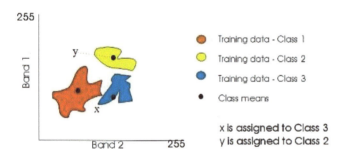

Figure 5.10 shows the same simplified example where there are only two bands of input data. The black dots represent the signature means. In this example, point x is assigned to class 3, even though it also quite close to class 1, because its distance to the mean of class 3 is smaller. Minimum distance to mean classification completely ignores the variability in a set of training data; it considers only the mean values. Thus, it can be quite sensitive to extreme individual pixel values (outliers), which will cause it to produce less accurate results.

Figure 5.10: *Minimum Distance to Mean classification*

Unlike **Boxcar**, the default minimum distance to mean procedure assigns every pixel to some class. This procedure does not produce any unclassified pixels. However, using the *Threshold* parameter, you can specify that the most deviant pixels be left unclassified.

Similarly, **MDM** does not produce mixed pixels. There may be overlap among signature ranges, but MDM unambiguously assigns pixels in the overlap area to one of the overlapping classes: the one whose mean is closest to that individual pixel.

To perform a minimum distance to mean classification, you must have a set of training signatures, either in a signature file or in memory. The signatures will be in memory if you have just finished executing **Classify⇒Training** or **Classify⇒Edit Signatures**.

MDM represents a type of multispectral classification: patterns of values across a set of spectral bands are used to recognize particular classes. Usually the larger the number of independent bands used in a classification, the better the results. (You can use **Utility⇒Scatterplot** to determine whether two bands

Copyright ©Global Software Institute

are independent.) You must use at least two bands. Note that you must use the same number of bands in your final classification as you used in defining your training signatures.

MDM requires you to select an image for each band requested. Normally these images should all represent the same scene. If you are using the same image bands that you used to define your training signatures, it is important that you name them in the same order as when you created the signatures, or your results will be erroneous. If you used a different scene for developing your training samples, you should still name the image files according to the same ordering of bands. That is, if you developed your training sets on Scene A, using Thematic mapper Bands 2, 3, and 5, and you now want to classify Scene B, you should use the Scene B files representing Bands 2, 3, and 5, in that order.

The **Minimum Distance to Mean** classification allows you to supply a threshold. For **MDM**, the threshold indicates a number of standard deviations away from the mean of closest class. The threshold is used to calculate a weighted Euclidean distance, using the standard deviations for each band of the closest class; if the spectral distance of a pixel from the mean of the closest class is larger than the calculated threshold distance, the pixel is considered to be 'unclassified', i.e. it is not assigned to any of the training classes. The larger the threshold value you enter, the fewer pixels that will be labeled 'unclassified'.

Once you have provided values for all parameters, **Dragon** reads the signatures from the signature file (if necessary) and then proceeds with the classification. When the classification is complete, **Dragon** displays a table showing the number of pixels in each class. The table is similar to that produced by **Boxcar** but does not show any confusion groups, since **MDM** does not produce mixed pixels. After showing the frequency table, **Dragon** displays the classified image, using a standard color scheme.

Note that even after classification is complete, the signatures used remain available in memory. Thus, you can save your classification results and then run a new classification using the same signatures but different image files.

5.8 Unsupervised Classification (UNS)

Choose one of the **Classify⇒Unsupervised** operations to classify a scene into categories based on its intrinsic structure, without using training signatures.

Clustering (CLU)

Choose **Classify⇒Unsupervised⇒Clustering** to classify a scene into categories by an iterative *isodata* procedure. No training signatures are needed for clustering an image. You simply choose the number of classes you want and the image bands you want to classify.

Dragon selects a set of starting cluster centers (*means*), based on the overall distribution of values in each band. Then each pixel is classified using a minimum distance to mean algorithm. After all pixels have been classified, the program reviews the results and recalculates the mean values for each class and band. Then the classification process repeats, using the new means to define cluster centers.

Over multiple iterations, the clusters should *converge*, that is, should become stable. Convergence will be indicated by the number of pixels that change class from one iteration to the next. When this number is relatively small, the clustering should stop since a relatively stable solution has been reached.

You may use up to four bands for clustering. You must specify an image for each band you use. You must also specify the maximum number and minimum number of clusters to be created (up to sixteen). If one of the clusters loses all its pixels, and this causes the number of clusters with non-zero frequency to fall below the minimum number you specified, a new cluster, with a different center, is started.

You have considerable control over the clustering process. Initially, you specify the (maximum) number of iterations as an operation parameter. At intervals, **Dragon** will give you the opportunity to stop the clustering. At these points, the system displays a table showing the number of pixels per cluster, the new cluster means, and the number of pixels that changed class membership during the last iteration, then put up a message box asking if you want to stop clustering.

The interval between questions is controlled by your response to the *Iterations: Between Questions* parameter. For example, if you specify a value of 3 for this parameter, **Dragon** asks after every third iteration whether you want to stop. If you do not want **Dragon** to question you at all, specify a value for this parameter that is larger than the maximum number of iterations you have requested.

If you click on the **Yes** button on the message box, or if all the repetitions specified have been completed, **Dragon** displays the usual class frequency table and then the clustered image.

You can save your clustering results using **Utility⇒Save**, and and assign labels to each cluster using **Utility⇒Header**. To modify the colors of each cluster, select **Utility⇒Colors**.

Image Slicing (SLI)

Choose **Classify⇒Unsupervised⇒Slicing** to classify a scene into categories defined by ranges of spectral values. No training signatures are needed for slicing an image. You simply choose the number of classes ("slices") you want, and **Dragon** determines the upper and lower bounds for each slice. Then **Dragon** classifies each pixel based on the range (the slice) in which its value lies, displays a table showing the number of pixels in each category, and displays the sliced image in the selected viewport.

You use only one band for slicing. You must specify an image for the band you want to slice, and a number of slices. The maximum number of classes (slices) that **Dragon** can handle is sixteen.

When slicing calculations are complete, **Dragon** will display a frequency table showing how many pixels fell into each slice. In slicing there can be no mixed pixels (since slices are defined to be non-overlapping). Also, there should be no 'unclassified' pixels. The slicing operation is designed to produce slices with about the same number of pixels in each (that is, the population of pixels is divided evenly, rather than the value range.) However, images will still show some irregularity in the pixel counts in each slice. For images with a very uneven distribution of data values, you may even find some slices containing no pixels at all.

After displaying the frequency table, **Dragon** will show the sliced image in the selected viewport. Since some slices may contain very few pixels, they may not be very visible.

Note
If you want to slice an image using specific cut-off points for each class, use the **Classify⇒Recoding** operation with a recoding table.

Agroclimatic Assessment (AGC)

Choose **Classify⇒Unsupervised⇒Agroclimatic Assessment** (commonly referred to as **AGC**) to recode and compress three NOAA Global Area Coverage (GAC) *channel files* into ACCS (Ambroziak Color Coordinate System) format. The operation automatically displays the resulting ACCS image using a special color scheme designed to mimic the Ambroziak H-I-S scheme.

The operation has three required parameters, the names of the three image files to be combined. (These must have been converted into **Dragon** format before you run **AGC**.) For correct results, the files must be entered in the order: Channel 1, Channel 2, Channel 4.

In addition to the required parameters, you can also enter a cloud threshold value for band 4, and scaling factors for Bands (channels) 1 and 2.

The default cloud threshold value is 90. Any pixels with Band 4 values greater than the cloud threshold are considered to be clouds. The default scaling factor is 100 for both Band 1 and Band 2. This factor produces no change in the raw data values. (Note the difference from conversion software developed by NOAA/NESDIS, where a value of 1000 is the default, no-change value. A scaling factor of, for example, 1050 in the NOAA system should be entered as 105 in **Dragon**.)

Like other **Dragon** operations, **AGC** leaves its results in image memory. To save the results, use **File⇒Save**.

Once you have saved a set of **AGC** results, you can use these results in any other **Dragon** operation. In particular, you may wish to re-display the ACCS-coded image, possibly to compare it to some other image or to use for a background in **Utility⇒Cursor**. The image created by **AGC** has image type *C*, and has its default color file set to use the special Ambroziak color scheme.

Because **Dragon** is a general-purpose image processing system, it provides many different possibilities for further analysis of **AGC** results. For example:

- You can display two saved ACCS images, from different seasons or years, side by side for visual comparison, using the two viewports.

- You can create a difference image to isolate areas of change between two ACCS images using **Enhance⇒Difference** or **Enhance⇒Symmetric Difference**.

- You can calculate a **Enhance⇒Normalized Vegetation Index** (NVI) image, a **Enhance⇒Global Vegetation Index** (GVI) image, or a simple **Enhance⇒Ratio** image using Bands 1 and 2, to examine the 'green-ness' component of the data separate from the brightness and cloud information.

- Using an ACCS or NVI image as background, you can examine individual data values from your channel data files, and/or from the ACCS/NVI files using **Utility⇒Cursor**.

- You can annotate your image with text using **Display⇒Annotate** operations.

- If you have a raster image that represents political boundaries, you can visually superimpose this on your ACCS image (**Display⇒Overlay**). Note that your boundary image should use a value of zero for non-boundary pixels, so that **Display⇒Overlay** will treat the non-boundary pixels as 'transparent'.

5.9   Recode Classes (REC)

Choose **Classify⇒Recode** to renumber classes or to combine several classes in a classified image, or to reassign or remap data values in any image. To use **Recode**, you must select one of several methods for specifying the old and new image data values.

The most straightforward method to use **Recode** is to provide four parameters: the image to be recoded, a range of values (first and last old class value) to be recoded, and the new value to be assigned to the range (first old value through last old value). When using this method, each selection of **Recode** can create only one new value, although a range of old values can be recoded. To make multiple reassignments, execute **Recode** several times successively.

You can also **Recoding** to apply a stretch or remapping to the selected image, using *Linear, Inverse, Gaussian*, or *Equalization* stretches. For convenience, you also have a choice of applying the same stretch used when this image was last displayed. (Note that, by contrast, the stretch applied while displaying an image does not actually modify the image data; it only affects the appearance on the screen.)

Finally, **Recode** can read a remapping table from an ASCII *table file* and apply that remapping to the selected image. The remapping table lists pairs of values; all pixels with the first value are changed to have the second value. The table can also recode a range of values to a single value. You can use this capability to do multi-range recoding of classified images, to implement piece-wise linear stretches, and for many other purposes. (The format of these files is discussed in detail in the *File formats* chapter of the *User Manual*.)

Recode also includes a mask image field. However, masking behaves somewhat differently in **Recode** than in most other operations. Areas that are zero in the mask image are not recoded; they are simply left with their old image values. This allows you to do recoding contingent upon other information. For example, you can recode all the water pixels to grass, **only** in regions where the predominant class is already grass.

When you have provided these parameters, **Dragon** will search though the classified image, replacing class values with the new values, according to your specifications. **Dragon** will display a message noting how many pixels' values have been changed. Then it will display the recoded image in the selected viewport. Note that the color scheme that you created for the original classified image will be used.

Since **Recode** leaves its results in memory, you must use **Utility⇒Save** if you want to make them permanent. After you save the recoded image, you can modify the color scheme using **Utility⇒Colors**.

5.10 Error Analysis (ERR)

Choose **Classify**⇒**Error analysis** to evaluate the accuracy of a classification. This operation compares the assigned class values in the image whose accuracy is to be evaluated with true class values, as indicated in a test data image, and accumulates a confusion matrix that shows how the test pixels were classified.

A confusion matrix provides detailed information on classification errors. The rows in the matrix correspond to actual classes, according to the test data. The columns represent the class values assigned by the classification procedure. The cells of the matrix contain pixel counts for each combination of actual and assigned class values.

Confusion Matrix

Assigned Class	Water	Vegetation	Bare	Unclassified	Total
True Class					
Water	2192	35	31	506	2764
Vegetation	0	1855	53	250	2158
Bare	0	25	2276	518	2819
Total	2192	1915	2360	1274	7741

	Error Statistics Summary			User Accuracy		Producer Accuracy	
Class Names	[True Class Pixels]	[Assigned to Class]	[Correctly Assigned]	[Commission Errors]	[Percent Correct]	[Omission Errors]	[Percent Correct]
Water	2764	2192	2192	0	100.00	572	79.31
Vegetation	2158	1915	1855	60	96.87	303	85.96
Bare	2819	2360	2276	84	96.44	543	80.74

Average Accuracy **81.98%**

Overall Accuracy **81.68%**

Kappa Coefficient **0.75**

Figure 5.11: *Classification error analysis result*

The diagonal cells of a confusion matrix show the number of pixels correctly classified. In a perfectly accurate classification, each diagonal cell would hold the total number of test pixels of the corresponding class, and all non-diagonal cells would hold zeros. In practice, this sort of result is rarely obtained, however, and at least some of the off-diagonal elements will be non-zero, i.e., will represent classification errors.

There are two categories of classification errors: errors of omission, and errors of commission. An error of omission occurs when a pixel that actually belongs to Class A is mistakenly classified as belonging to some other class. An error of commission occurs when a pixel which actually belongs to Class B is assigned to Class A instead. Labeling a misclassified pixel as an error of omission or commission is always relative to a particular class. If there were only two class labels available, A and B, all errors of omission with respect to Class A would be errors of commission with respect to Class B, and vice versa.

Classify⇒Error analysis will display the confusion matrix in a report window. Figure 5.11 shows an example. You can save the report as an HTML file. The report includes a table of summary statistics, showing correct classifications, errors of omission and errors of commission for each class, as well as the average accuracy and overall accuracy. The latter two measures consider only errors of omission. Average accuracy is computed by summing the individual percent correct values for each class and dividing the total by the number of classes. Overall accuracy is calculated by multiplying each percent correct by the number of test pixels for the corresponding class, summing these weighted percents, and dividing by the total number of test pixels (over all classes).

Finally, the error analysis operation calculates and displays the *Kappa coefficient* for the confusion matrix. This statistic has been recommended in the remote sensing literature as a suitable accuracy measure for thematic classification, since it considers all elements of the confusion matrix. Basically, Kappa measures how far the distribution of values in the confusion matrix deviates from a chance arrangement. The larger the value of Kappa, the higher the accuracy of the classification. A perfectly accurate classification would produce a Kappa value of 1.00.

Classification error analysis requires two parameters. The first is the name of the classified **Dragon** image file to be evaluated. If a non-classified image is provided, the operation will produce nonsensical results.

The second parameter is the name of the **Dragon** image file that holds the test data. This file, called the 'Class Map image file' in **Dragon**, can be created in several ways. One method is to save the contents of image memory into an image file after doing classification training via **Classify⇒Training⇒Define** or **Classify⇒Training⇒Apply**. Another method for creating the test data file is to define and fill polygons at the test data locations, in the **Geography⇒Polygons⇒Vector** operation, and then save the rasterized results in an image file.

It is possible to use the same pixels for training and for evaluating accuracy. However, this practice will produce biased accuracy assessments, typically resulting in higher accuracy levels than are justified. Ideally, training and test pixels should be selected independently.

Regardless of how it is created, the class map file will normally have zero values in most positions. At the row/column positions corresponding to test pixels, it should hold the true class at that location.

5.11 Exercises

This section offers a set of hands-on exercises with some of the **Dragon** classification operations, in order to let you practice using these functions and see their results.

1. Start your **Dragon** system. Choose **Display⇒Gray** Enter the following parameters:

   - Image File: mek14-3.img
   - Viewport: 0

   Click the **OK** button to start the operation. **Dragon** displays the image, a near IR band from the SPOT 4 satellite, showing the Mekong River near Nongkhai, Thailand during the dry season. We will use thresholding to identify water versus non-water areas in this image.

   Choose **Classify⇒Recode**, enter the following parameters, and then click **OK**:

   - Image File: =M
   - Recoding method: Values
   - First Old Value: 0
   - Second Old Value: 48
   - New Value: 1
   - Display option: Gray
   - Viewport: 1

   This sets all values from 0 through 48 (our threshold) to class 1. You will not see much change in the image since these pixels were already dark.

Choose **Classify⇒Recode** again. Enter the following parameters, and then click **OK**:

- Image File: =M
- Recoding method: Values
- First Old Value: 49
- Second Old Value: 255
- New Value: 2
- Display option: Color
- Viewport: 1

This sets the remaining pixels, the ones above our threshold to class 2. The result is an image with only two values. To see the image in color, choose **Display⇒1 Band**, select *Retrieve* for the *Color file option* parameter, then select the color file **defcls.clf**.

2. Choose **Display⇒3 Band** Enter the following parameters and then click **OK**:

- Image File for Blue: mek14-1.img
- Image File for Green: mek14-2.img
- Image File for Red: mek14-3.img
- Viewport: 0

Dragon displays a three-band color composite of the Mekong River image. The red areas are vegetation. Notice the extensive regions of bare soil (tan/brown) and the white sandbar in the river. We will use this image as the background to interactively select training areas and calculate signatures.

3. Choose **Classify⇒Training⇒Define**. Enter the following parameters and then click **OK**:

- Background Image File: =C
- Band 1 Image File: mek14-1.img
- Band 2 Image File: mek14-2.img
- Band 3 Image File: mek14-3.img

Dragon displays the color composite, plus a panel of buttons for controlling the process of training. Click on the **New Class** button or press **<F6>**. Enter the signature name "River" in the dialog that pops up and click **OK**.

Now click on a point in the middle of the river. (If you are using **Dragon Professional**, you must select a region first.) **Dragon** draws an "X". Now click again, a few pixels away, but still in the river area. **Dragon** draws a circle around the first selected point, with a radius by your second point.

Look carefully at the circle. Does it contain only pixels from the river (greenish blue)? Or does it include some pixels from the land? If you see any pixels in the circle that are not water, click on the **Delete** button or press **<F9>** and try again. If this river area looks okay to you, choose another circle in a different part of the river. Be sure to pick only water. Don't include the sandbar.

Click on the **New Class** button or press **<F6>**, then enter the signature name "Sandbar". Select one or two training areas in the white area along the bank of the river. Once again, be careful that you do not include pixels from other classes in your circles. If you do, delete the circle using **<F9>** and try again.

Click on the **New Class** button and enter the signature name "Vegetation". Select several training areas in the reddish areas of the image.

Click on the **New Class** button and enter the signature name "Bare". Select several training areas in the brown or tan areas of the image. Select at least one area on each side of the river.

Now click on the **Accept** button or press **<F1>**. This is a signal to tell **Dragon** that you are finished defining training areas and that you want to calculate signature statistics. **Dragon** locates pixels from the training areas you defined in the three input images and calculates the mean, standard deviation, minimum and maximum values, and covariance matrix for each class. The results of these calculations remain in memory where you can view or save them. Meanwhile, **Dragon** displays an image showing you the locations of each training area, using different colors for different classes. You can save this image and use it as a class map image in the **Classify⇒Training⇒Apply** or **Classify⇒Error Analysis** operations.

4. Choose **Classify⇒Edit Signatures⇒Show**. Enter the following parameters and then click **OK**:

- Signature File: (Current)
- Name of Signature to show: River

Note that you must tab off the *Signature File* field before the *Signature Name...* list will be populated.

Dragon shows all the values it calculated for the "River" signature. Choose **Classify⇒Edit Signatures⇒Show** again and look at the remaining signatures. If you choose the "*" value for *Signature Name*, **Dragon** will show all the signatures.

5. Choose **Classify⇒Edit Signatures⇒Histogram**. Enter the following parameters and then click **OK**:

- Signature File: (Current)
- Name of Signature to display: Vegetation

Dragon displays the histogram of the data in all three bands of the signature. A good signature should have a fairly narrow histogram with a single peak, in all bands. Examine the histograms for the other signatures you created. If your histograms look too broad, or have multiple peaks, you might want to repeat the training step, being more careful in selecting your training areas.

6. Choose **Classify⇒Edit Signatures⇒Save**. Enter the following parameters and then click **OK**:

- Signature File for save: mekong.sig

Now choose **Classify⇒Supervised⇒Boxcar**. Enter the following parameters and then click **OK**:

- Signature File: mekong.sig
- Band 1 Image File: mek14-1.img
- Band 2 Image File: mek14-2.img
- Band 3 Image File: mek14-3.img
- Viewport: 0

Dragon uses the boxcar algorithm to assign a class to each of the pixels in the image. It displays the resulting image, with each class coded in a different color, and a table showing the number of pixels assigned to each class.

As explained in Section 5.3, the boxcar algorithm can produce "mixed pixels", that is, pixels whose class membership is ambiguous. Do you have many mixed pixels your classification results? Check the table and the image. Mixed pixels show up in a neutral gray color in the classified results. Which classes are most frequently confused, according to the table?

The boxcar algorithm can also result in unclassified pixels. Do you have any unclassified pixels in your results? Check the top line of the results table, or look for black areas in the resulting image.

7. Choose **Classify⇒Supervised⇒Maximum Likelihood**. Enter the following parameters and then click **OK**:

- Signature File: mekong.sig
- Band 1 Image File: mek14-1.img
- Band 2 Image File: mek14-2.img
- Band 3 Image File: mek14-3.img
- Viewport: 1

Dragon uses the maximum likelihood algorithm to assign a class to each of the pixels in the image. It displays the resulting image, with each class coded in a different color in Viewport 1, plus a table showing the number of pixels assigned to each class.

Maximum likelihood does not produce mixed pixels. Normally it will not produce "unclassified pixels" either, but **Dragon** implements a threshold function that sets extremely different pixels to be unclassified. Does your classification result have any unclassified pixels? Which result do you think is better, the maximum likelihood or the boxcar?

Choose **Display⇒Annotation⇒Legend** and place a legend on the image in Viewport 1. This will let you see which classes are being displayed in which colors (although you can probably guess). If you don't like the colors used by default, you can create your own color file using the **Utility⇒Colors** operation.

8. Choose **Classify⇒Unsupervised⇒Clustering**. We will compare the results of unsupervised classification with your maximum likelihood results. Enter the following parameters and then click **OK**:

   - Band 1 Image File: mek14-1.img
   - Band 2 Image File: mek14-2.img
   - Band 3 Image File: mek14-3.img
   - Minimum number of clusters: 4
   - Maximum number of clusters: 4
   - Number of iterations: 20
   - Number of iterations between questions: 20
   - Viewport: 0

Dragon runs the clustering process twenty times (which can take a few minutes) and then displays a table plus the results in image form. How closely do the clustering results match the supervised classification? (Remember that clustering assigns class numbers arbitrarily, so the class numbers and colors will not necessarily be the same. You can use the **Classify⇒Recode** operation if you want to change the clustering results to use compatible class numbers.)

Clustering will never produce unclassified pixels. Every pixel will be assigned to some class. Try the clustering again, setting the minimum and maximum number of classes to 8 and using Viewport 1. Does allowing more classes appear to improve the result?

Chapter 6

GEOGRAPHY OPERATIONS

6.1 Principles of Geometric Correction and Raster GIS

The **Geography** menu groups together various operations that require the user to interact with the image in a viewport: by drawing, measuring, extracting areas, selecting points, and the like. These operations have been grouped together because they share many similarities in their mode of use, and in some restrictions as well. However, they serve a variety of different objectives.

Geography also includes operations that would commonly be included in a raster-based Geographic Information System (GIS).

The capabilities in **Geography** are so diverse that it is not feasible to introduce them with a single set of general principles. Furthermore, the sections in this chapter that describe individual operation provide considerable background on the purpose and logic of each one. Therefore, we concentrate on providing background for the two largest topics handled via **Geography**: geometric correction/registration and raster GIS operations.

Geometric Correction

Images captured by remote sensing instruments provide information about phenomenon at particular locations on the ground. However, as explained in Chapter 1, raw remote sensing imagery contains distortions and errors. These distortions mean that initially, it is not possible to accurately identify the exact geographic position associated with each pixel. To make a remotely sensed image useful for *quanitative analysis* (that is, extraction of information about distances or areas), the image must be transformed to match some geographic coordinate system such as UTM. This process of matching an image to geographic coordinates is called *geometric correction* ("geocorrection" for short) or *image rectification.*

Geometric correction is a mathematical process that has three steps:

1. **Determine the mathematical relationship between the coordinate system of the raw image (pixel coordinates) and the desired reference coordinate system (e.g. UTM).** This relationship will usually be expressed as an equation relating reference coordinates to the original coordinates. The form of this equation can vary. In **Dragon** we use a relatively simple form. More complex transformations might allow differential modifications of different parts of the image. These complex transformations could result in more accurate results but are far more difficult to understand as well as to compute. **Dragon** determines an appropriate equation by looking at pairs of locations in the original and reference locations (called *control points*) and applying *linear regression*. Linear regression tries to fit a linear equation (that is, an equation of the form $x = ax + by + c$) to that predicts the value of one set of data from the other.

2. **For each pixel in the output image, determine the corresponding location in the source image.** This is accomplished by applying the linear equation inferred in step 1. Often the source location will not map to exactly one pixel position.

3. **Calculate the appropriate value for the output pixel.** If the position computed in step 2 maps to exactly one pixel in the source image, this step is trivial. Usually, however, the output pixel value will need to be based on the values of several source pixels. There are several methods for combining the values from source pixels to produce an output pixel value. **Dragon** provides two: *nearest neighbor* and *bilinear interpolation*, which are described in detail in later sections.

In addition to geometrically transforming the original image to match a reference coordinate system, geometric correction can also change the scale, that is, the size of one pixel on the ground. This process is called *resampling*. **Dragon** allows you to specify the size of the output pixel in order to support resampling.

Figure 6.1 illustrates the basic concept of geometric correction.

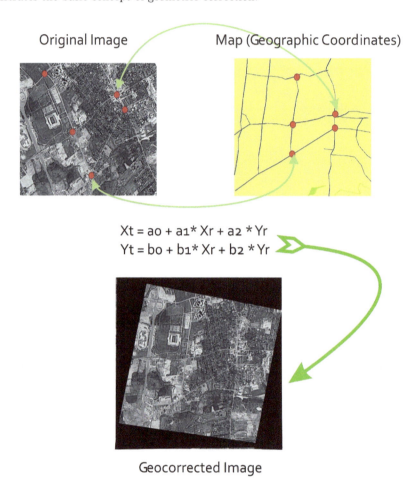

$$Xt = a_0 + a_1 * Xr + a_2 * Yr$$
$$Yt = b_0 + b_1 * Xr + b_2 * Yr$$

Figure 6.1: *Geometric correction concepts*

Images that have been corrected to match a geographic coordinate system are said to be *georeferenced*. This means that it is possible to state the geographic coordinates of any pixel in the image. When you geocorrect an image in **Dragon**, the system stores information about the geographic coordinate system in the image metadata. This allows **Dragon** to display geographic pixel coordinates in the **Utility⇒Cursor** operation and measure lengths, perimeters and areas in the **Geography⇒Measure** operation (both of which are described below).

Note that in some cases, images acquired from data providers will have already been geocorrected. In this case, **Dragon** will read and store the georeferencing metadata as part of the process of importing the images.

For maximum flexibility, **Dragon** separates the different steps involved in geometric correction into different operations. One operation is used to choose control points by picking them out on the source image. A second operation computes the linear equation for transformation, allowing the user to specify the desired accuracy. A third operation actually performs the transformation and resampling to produce an output image that matches the desired coordinate system.

Up to this point, we have assumed that the goal of geometric correction is to transform the source image to match coordinates on a map. In addition, there are some situations where we might want to geometrically transform an image to match a second image. This is usually called *registration*. The reference image may or may not have been transformed to match some map coordinate system. The goal of registration is to geometrically align two images from different times or sources.

Image-to-image registration is particularly critical for *multi-temporal image analysis*. Frequently we will want to know how some phenomenon has changed over time. For example we might want to measure how urbanization has increased over several decades, or to evaluate beach erosion before and after the creation of a breakwater. Multi-temporal analysis requires image data from multiple dates covering the relevant period of time. Sometimes the images might come from different sources; older images might be aerial photographs, for example, while more recent images might be satellite-based. The different images may have different scales as well. Image-to-image registration transforms all the images into the same geographic frame of reference so that they can be combined (using operations like **Enhance⇒Sum** or **Enhance⇒Difference**) and compared.

Although the goals of image-to-image registration are different from geometric correction, the mathematical process is exactly the same. The user defines pairs of points in the source and the reference image. The image analysis system then computes the relationship between the two images and transforms the source image to be geometrically compatible with the reference image. The resulting transformed image may or may not be georeferenced, depending on whether the reference image was previously geocorrected or not.

Raster GIS

As explained in Chapter 1, a raster is a regular grid or matrix, consisting of equal-sized cells. When a raster is used to represent geographic phenomena, each cell is associated with a specific area on the ground. Individual cells can be referenced by their column and row (x and y) coordinates. (A raster may also be georeferenced, so that cells can be designated by their geographic coordinates as well.) The numeric value associated with a cell represents some measure or numerically-encoded property of the corresponding area on the ground.

Digital remotely-sensed images are almost always represented in raster form. Each cell is a pixel. The value in each cell is the reflected or emitted radiation measured by the sensor from that area on the ground.

However, a regular grid of cells can be used to represent many other kinds of geospatial information, aside from what can be measured directly via remote sensing. One important example is the elevation (height) of the ground compared to sea level. A raster holding elevation values is called a *Digital Elevation Model* (DEM). DEMs can be used to calculate the *slope* (steepness) of a surface as well as its *aspect* (the direction it faces). Both basic elevation and the derived quantities of slope and aspect are important considerations in solving problems such as where to site a building, where to plow and plant crops, or where to construct a road. DEMs can also be used to predict where water will flow and how fast, so they are important in applications such as water resource management and flood prediction.

Other categories of geographic information that might be represented in raster form include measured rainfall and temperature, soil type, vegetation type, wildlife sightings, population density, and so on. Raster representations are particularly appropriate for geospatial quantities that change fairly gradually over space. They are less useful for quantities that have sharp boundaries, such as political entities, building footprints, or water bodies. Although they can be used for such quantities, the regular spacing of the cells and the fact that each cell can hold only a single value will introduce errors in edge areas.

A *raster GIS* is a system that analyzes spatially-distributed phenomena that are represented in raster form, in order to answer questions or make decisions. A raster GIS manipulates information in *layers*. Each layer represents a different quantity, distributed over an area of interest. A specific cell (at a specific column/row) represents the same location on the ground across all layers. Thus raster GIS layers are similar in concept to the bands of a multispectral image.

However, layers may come from many sources. Some, for example, vegetation type, may result from classification operations performed on remote sensing imagery. Elevation layers are often created from *stereoscopic imagery*, that is, paired images of the same area captured from different angles. Layers such as population density or land ownership may be created by importing information from spreadsheets or databases. Very often, critical layers will be created by performing computations on other layers. Slope and aspect layers can be derived from an elevation layer. A population growth rate layer could be created by taking the difference between two population density layers representing different time periods.

Operations in a raster GIS create new layers by performing computations on one or more existing layers. Computations may involve a single layer, such as slope calculation, or may combine multiple layers. Layer combination operations may be arithmetic or logical. Arithmetic combinations perform some set of calculations on the values in corresponding cells from the input layers. Many **Dragon Enhancement** operations can be used for arithmetic layer combinations as well as for enhancing remote sensing images. The **Classify⇒Recode** operation and the **Geography⇒Combine Layers** operation are logical operations. In particular, **Geography⇒Combine Layers** can produce an output layer that is a logical combination of as many as twelve input layers.

The concepts and basic operations of raster GIS are quite simple. Understanding how to apply these simple building blocks to get the desired result is far more challenging. One helpful way to view raster GIS is as a series of steps, each of which applies some operation to some input layers and produces one or more output layers. This sort of input/output diagram is sometimes called a *GIS model*.

Figure 6.2 shows an example of a simple GIS model. The problem in this example is to find the best location for creating a new picnic area along a mountain highway. The criteria for selection are as follows:

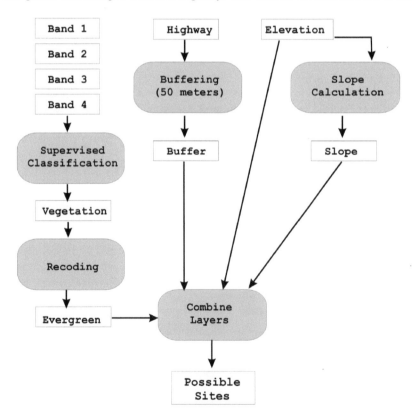

Figure 6.2: *Simple GIS model*

1. The picnic area site must be within fifty meters of the highway.

2. The site must have an elevation of more than 1000 meters, to insure that it will have a good view.

3. The site must have a slope no greater than 3% (that is, it must be close to flat) so cars can easily park there.

4. The site must be in an area with evergreen trees, for shade and scenic value.

Assume that when we begin the analysis, we have four georeferenced bands of remote sensing imagery with 10 meter per pixel resolution, plus a DEM with 10 meter resolution and a raster layer of the same resultion showing the location of the highway. (This highway could be created from a vector representation of the highway using the **Dragon Geography⇒Vector** operation.)

First, we use supervised classification to create a vegetation type layer. Since our criteria require only that we distinguish evergreen trees from other types of vegetation, the classification does not need to include many classes. We use the **Classify⇒Recode** operation to change the evergreen class to value 1 and all the other classes to 0.

Next, we use the **Geography⇒Buffer** operation. This operation calculates distances from cells that have a specific target value. The input is the highway layer. We identify all cells that are fifty meters or less from a highway cell and assign those the value of 1. All other cells receive a value of 0.

We calculate the slope of each cell using **Geography⇒Slope**. We assign a value of 1 to all cells with a slope of 3% or less.

Finally we are ready to logically combine all our layers using rule-based logical combination (**Dragon Geography⇒Combine Layers** operation). Our rule says:

```
if EVERGREEN=1 and BUFFER=1 and SLOPE=1 and ELEVATION>1000
then set output to 255 otherwise set output to 0
```

This rule is the GIS equivalent of our selection criteria. The result of this operation will be a new layer which has the value 255 in every cell that satisfies our criteria and is thus a potential picnic area site, and a 0 in every other cell. We can visualize the results by using the **Dragon Display⇒Overlay** operation to show the potential sites "on top of" a three-band composite of the remote sensing image.

How can you design a GIS model that will give you the results you need? One approach is to start with your goal. What input layers would you need in order to execute the final step in your process (which will often be a multi-layer logical combination)? If you don't already have those layers, what would you need create them? By working backwards, you can define a chain of operations that leads from the original input data you have available to your final objective. Sometimes this strategy will reveal that you need additional primary data layers, layers that cannot be derived from what you already have. If you began with your available input layers and worked forward, you might not discover this requirement.

6.2 Overview

Geography consists of three categories of operations. In the first category (in a submenu called **Geography⇒Geometry**) are operations used to transform an image so that it is registered to another image or to a base map. These operations are particularly important if you plan to use the results of image processing operations in conjunction with other geographical data (for instance, in a GIS). They are also essential if you intend to use multi-sensor or multi-temporal analyses.

The second category of operations (in a submenu called **Geography⇒Polygons**) allows you to draw lines and polygons on top of an image. For example, **Geography⇒Polygons⇒Vector** allows you to create an 'image' that holds only lines and filled or unfilled polygons. This image might, for example, show locations of rivers and water bodies in a geographical region. This kind of image can be saved in a Layer file and then can be displayed on top of a normal image file (using **Display⇒Overlay**), combined with other images (using **Enhance⇒Mask**, **Enhance⇒Difference**, etc.), used to restrict processing in operations which accept a *Mask File* parameter, or used in GIS analysis. The other **Geography⇒Polygons** operations allow you to measure lines and areas, using an image as a background, and to extract an irregularly shaped subset from an image.

The third category of operations, in the **Geography⇒Raster modeling** submenu, provides facilities for creating or combining raster data layers that represent information other than reflected electromagnetic radiation. These include the ability to calculate slope or aspect layers from an elevation layer, the ability to identify areas within a specific distance of some target feature or class, and the ability to combine a set of input layers to create an output layer, based on conditional rules.

6.3   Geography Interaction

Many **Geography** operations are highly interactive. You control the positioning of the cursor and indicate when to select a new point on a line, close a polygon, fill a polygon with a solid color, and so on. The 'line drawing' **Geography** operations all use very similar procedures. **Geography⇒Geometry⇒Ground Control Point** selection is somewhat different but has the same basic flavor.

Generally, the interactive **Geography** operations are used as follows:

1. You must specify a background image through response panel input or command parameter specifiers. In many cases, you can specify the composite image (=**C**) as background.

2. **Dragon** reads (if necessary) the image file you specified, and displays it.

3. If you are using **OpenDragon**, you can begin choosing points immediately. However, in **Dragon Professional** the image will initially be displayed as an *overview*. You may choose points within the overview, or you may choose a *region of interest* using **<F11>**, as described in *Chapter 2*. The region of interest will then be displayed at a 100% (1-to-1) resolution. You may return to the overview (using the right mouse button) and select a new region of interest as often as needed during the interaction.

Note
Ground control points can **only** be selected within the region of interest, not within the overview. You must select a region before starting to choose points.

4. Move the mouse cursor to a location that you want to select. You can use either the cursor keys or a mouse for these operations. To select a point, you either press the left mouse button or the **<Ins>** (insert) key.

5. Move the cursor to the next desired location. As the cursor moves, **Dragon** displays a 'rubberband' line from the last selected point to the current cursor position. When you select a point, the line is anchored and a new line starts at the selected point.

6. By moving the cursor and selecting points, you can draw a multi-segment line or polygon. To signal that you want to complete the current figure, you click on **End Line** (or press **<F6>**) to finish a polyline or click on **Close Polygon** (or press **<F7>**) to close and finish a polygon. Note that **Close Polygon** will automatically draw the final line segment from your last-selected point to the starting point of the polygon.

When you have completed a line or polygon, the different **Geography** operations offer different actions that you can perform on that figure. These actions can be selected by clicking on the appropriate buttons, or pressing function keys. For instance, you must click on **Fill Polygon** (or press **<F8>**) in **Geography⇒Polygons⇒Vector** to fill the last-constructed polygon with a solid color.

All of the interactive **Geography** operations use the **<F1>**, or the top button on the button panel, key as a signal to terminate the interaction, do final processing and exit the operation. For instance, in **Geography⇒Polygons⇒Vector**, clicking on **Rasterize** (or pressing **<F1>**) will rasterize the lines and polygons you have specified to create a data layer in image memory. In most cases, the **Cancel** button, or the **<F10>** key, terminates the interaction without doing final processing or modifying image memory.

Geography⇒Polygons⇒Measure is an exception. In this operation, **<F10>** is not active and there is no **Cancel** button, since there is no final processing to be done.

All of the interactive **Geography** operations allow you to zoom the image. This makes it easier to select specific locations on the image. To zoom the image, use the **Zoom** menu on the viewport. If you are using **Dragon Professional** the **Zoom** menu will only be enabled after you have selected a *region of interest*.

6.4   Files Used by Geography Operations

The Geography operations create and use some special types of files not relevant to other **Dragon** operations. These files are discussed briefly below. Their structure is detailed in the *File Formats* chapter of the *User Manual*.

Layer files: Coded image files which are created by saving the contents of image memory after executing **Geography⇒Polygons⇒Vector**. The structure of layer files is discussed in *Chapter 2*. **Dragon** will not normally apply a contrast stretch to a layer file. Also, layer files produced by **Vector** frequently include large background areas with a value of zero. Hence they are ideal for use as overlay images in **Display⇒Overlay**, or as masks. Sometimes when we are discussing coded image files, we use the term *cell* as opposed to *pixel*. The latter term implies that the value reflects some physical measurement; the former is more neutral, recognizing that a coded image can be an array of any type of values.

Ground control point (.gcp) files: The **Geography⇒Geometry⇒Ground Control Point** operation creates these files, which are used by **Geography⇒Geometry⇒Calculate** (calculate registration coefficients). A GCP file holds a series of text labels followed by x and y coordinate values. **Calculate** requires two GCP files: a *target* GCP file, which holds coordinates for control points extracted from the image to be registered, and a *reference* GCP file, which holds control point coordinates from the map or the base image that defines the 'standard' coordinate system. To calculate the equations for transforming a target image file to a reference coordinate system, **Calculate** needs pairs of corresponding target and reference coordinates. It assumes that points with matching labels in the target and reference GCP files correspond.

Coefficient (.cof) files: Calculate produces a coefficient or *parameter* file that summarizes the mathematical relationship between the target and reference coordinate systems. This coefficient file is required input for the **Register** operation, which is used to geometrically-correct or register an image.

Calculation Output (.out) file: Calculate is a cyclic computation. Rather than clutter the screen with the ongoing results of this process, **Calculate** writes its processing to a file. You can then examine or print this file using normal operating system capabilities.

Measurement data file: Measure can create a data file that holds values from line profile and area histogram operations. This file can have any file name. It is a text file with a structure that is intended to be easy to edit or import into databases, spreadsheets, and graphing packages. See the *File Formats* chapter in the *User Manual* for further information.

Rule file: Geography⇒Raster⇒Combine Layers uses a special file that specifies what output value to assign to a cell in the result based on the the values of the corresponding cell in the input images or layers. Rule files are created in the **Geography⇒Raster⇒Create Rule**. They can also be created or edited manually. A rule file is a text file with a structure that is intended to be easy to read and edit.

6.5   Geometry Operations

Choose **Geography⇒Geometry** to gather control points and register images.

Ground Control Point Selection (GCP)

Choose **Geography⇒Geometry⇒Ground Control Point** (usually known by its abbreviation **GCP**) as a first step toward geometrically-correcting an image or registering one image to another. In **GCP**, you select a set of readily-identifiable locations on the image you wish to register or correct. **Dragon** saves the coordinates of these points, along with identifying labels that you provide, in a Ground Control Point (GCP) file.

The control points you select from your target image must be matched with a corresponding set of control points from the reference coordinate system. These may be either map coordinates (e.g. UTM coordinates) or image coordinates (i.e. pixel and line numbers). The control points in the reference GCP set are assumed to represent the same locations as the correspondingly-labeled target control points. Given this assumption, **Dragon** can compute the mathematical transformation needed to convert the target image into the reference coordinate system. (This is the function of **Geography⇒Geometry⇒Calculate**.)

You can use **GCP** to record reference as well as target control point coordinates, depending on the parameter values you enter. Alternatively, you can use a reference GCP file created through some other process (e.g. by an external digitizing program) to guide the selection of target image control points.

GCP Parameters

The inputs required and the outputs produced by **GCP** are controlled largely by your response to the prompt *Source for Reference Coordinates*. There are four choices available for this parameter:

- If you specify *None*, **Dragon** will create a single GCP file holding image coordinates only, consisting of a label for each point plus the image line and pixel numbers.

- If you specify *Image*, **Dragon** will display a reference image side by side with your target image. You select corresponding control points on the target and reference images. **Dragon** creates two GCP files, with corresponding labels.

- If you specify *File*, you must also supply the name of a GCP file holding reference GCP coordinates. **Dragon** will prompt you to select control points based on the labels in this file. Only a single GCP file (with coordinates derived from the target image and labels derived from the reference file) will be created.

- If you specify *Keyboard*, then as you select control points on the target image, **Dragon** prompts you to type the reference coordinates for that control point. These may be (for example) coordinates read off a map. **Dragon** creates two GCP files, one holding coordinates of selected image points, the other holding the typed coordinates.

Most of the other **GCP** prompts are fairly self-explanatory. Several of the fields are disabled until you select some value for *Source for Reference Coordinates* which makes them necessary.

If you are using the *Image* method for selecting GCP points, the *Reference Image File* field will be enabled. You must provide the name of the image to which the *target* will be registered. The target image is displayed in Viewport 0 while the reference image is displayed in Viewport 1.

Also, if you are using the *Image* method, the *Output Reference Coordinate file* field will be enabled and is required.

If you are using the *File* method for selecting GCP points, **Dragon** will enable the *Input Reference Coordinate File* field. In this field, you must enter the name of the reference coordinate file which holds previously collected reference control points for the target image.

If you are using the *Keyboard* method for selecting GCP points, **Dragon** will enable the *Output Reference Coordinate File* field. The filename entered here will determine where the reference coordinates gathered from the keyboard will be stored.

Note
If you use =**M** as your response to the *Image File to be Registered* prompt, **Dragon** will provide a default target GCP file name of **target.gcp**. If you accept this value, any previous GCP file with this name will be overwritten. However, you can change this value if you wish.

GCP Interaction

Click on **Ok** after entering GCP parameters in the GCP response panel to begin the control point gathering process. The details of this process depend somewhat on the options you have selected, but the basic structure is as follows:

1. **Dragon** reads and displays the target image.

2. Depending on *Source* method, **Dragon** either reads and displays the reference image, or reads the reference coordinate file.

3. **Dragon** prompts you (via a message in the Viewport 0 status line), to enter a control point on the target image. To do this, position the mouse cursor and click with the left mouse button, or press the **<Ins>** key.

4. **Dragon** pops up a dialog to prompt you for a control point label. The default name for this point is **Pt#**, where '#' is replaced by the sequential control point number. You can enter any label up to 32 characters long without spaces. We recommend that you use descriptive labels (e. g. **I95-Route2**) that will assist you in locating the control point in different data sources. Labels in different GCP files must match exactly, except for case, in order to treat them as corresponding points.

   If you selected the *File* method for *Source for Reference Coordinates*, you can only enter labels that match labels in the reference coordinate file.

   If the label that you type is valid, **Dragon** will display it on the image at the position that you selected.

5. If you chose *Image* method for *Source for Reference Coordinates* **Dragon** prompts you to select a corresponding control point location on the reference image (in Viewport 1). **Dragon** labels the reference point after you select it.

Note

In **Dragon Professional**, you must select a region of interest in both the target and the reference images before you can pick points in either image.

6. If you chose the *Keyboard* method, **Dragon** pops up a dialog where you should enter the X and Y reference coordinates.

7. Go back to step 3 and repeat.

To exit from **GCP**, press the **<F1>** key or click on the **Accept** button instead of selecting a new control point on the target image in step 3. At this point, **Dragon** will write the gathered coordinate data to GCP file(s) and then exit the GCP processing. To exit from GCP without saving any control points, press the **<F10>** key or click on the **Cancel** button.

Dragon allows you to zoom the image before choosing a point, in order to be able to select a point more precisely.

You can select as many points as you want while the image is zoomed. To return to the unzoomed state, choose **Zoom⇒Restore**. The image will be redisplayed at its normal size, and previously selected control points will be redrawn.

Guidelines for Selecting Control Points

This section briefly summarizes some rules for selecting control points. Consult the references in the *Bibliography* or other image processing texts for more detailed guidelines.

The accuracy of your registration is directly affected by the accuracy of your control points. If target and reference control points do not represent exactly the same location, errors will result. Obviously the accuracy of selection is limited by the resolution of the data. Given this limitation, however, control point selection strategy can have a significant impact on accuracy.

Some common guidelines for control point selection are:

- Select control points from varied regions of the image. If you select all your control points within a relatively small area, you run the risk that the distortions present in that area are not representative of the image as a whole. Furthermore, quantizing errors in the selection process are magnified at regions of the image distant from the points chosen. Ideally, you should try to get some points from each quadrant of the image.

- Select points that are clearly identifiable in both the target and reference data sources.

- Select points that correspond to reproducible, 'point' features. Artificial features such as buildings, right-angled road intersections, corners of cleared areas, and so on, make good candidates. Geological features such as canyon heads or ridges can also be useful. Avoid using details of coastlines or other water bodies, as this information can change dramatically across seasons or over longer time spans.

- Select an adequate number of points. The minimum number of points to support the regression analysis in **Geography**⇒**Geometry**⇒**Calculate** is four. Normally you should have at least double this number, to allow for discarding of some points due to inaccuracy.

GCP Limits

GCP is subject to several limits, as follows:

- A maximum of 30 control points can be entered in a GCP file (and used in **Calculate**). **Dragon** warns you when you have entered 30 points, and then automatically saves your control points and exits from GCP.

- If you are using a pre-existing GCP file as as the *Source for Reference Coordinates*, you can enter only as many control points as are defined in that file. Also, all labels entered must match one of the labels in the reference file.

- Control point labels can be up to 32 characters long, and may not include spaces.

Calculate Registration Coefficients (CAL)

Choose **Geography**⇒**Geometry**⇒**Calculate** to calculate the transformation equations needed to register a target image to a reference coordinate system. **Calculate** produces a coefficient file that can be used as input to the **Geography**⇒**Geometry**⇒**Register** operation, to specify the parameters for registering or geometrically correcting an image.

This section briefly describes the computations and rationale of **Calculate**. For a more complete discussion of regression, coordinate transformations, error measures and other topics related to image registration, consult references in the *Bibliography* or other image processing texts.

Calculate requires two sets of ground control points, one derived from the target image and one representing the reference coordinate system (image or map). **Calculate** assumes that the target and the reference GCP files correspond; specifically, the operation assumes that points in the two files that have matching labels represent the same location in space. As a first step, **Calculate** matches the contents of the two files and reports on how many matching coordinate pairs it found. (Non-matching points are discarded.)

Once **Calculate** has extracted corresponding pairs of control point coordinates, the operation performs a regression analysis on these pairs of points, using the reference coordinates as independent variables to try and 'predict' the target coordinates. The regression analysis produces a set of equations that describe the mathematical relationship between the target and reference coordinate systems. The form of these equations is as follows:

```
Xt = a0 + (a1 * Xr) + (a2 * Yr)
Yt = b0 + (b1 * Xr) + (b2 * Yr)
```

Xt and Yt represent X and Y coordinates in the target coordinate system, while Xr and Yr represent the reference coordinate system. The six coefficients (a0, a1, a2, b0, b1, b2) are calculated by the regression process so as to minimize the overall discrepancy between the actual (Xt,Yt) coordinates of control points and the theoretical (Xt,Yt) calculated by plugging control point (Xr,Yr) pairs into these equations. These coefficients completely describe the transformation between the target and reference coordinate system.

A transformation with equations of this form is called an *affine transformation*. Other types of transformations are often used for geometric correction and registration. However, the affine form provides reasonably accurate results for moderate-sized image subsets, and has the advantage of calculational simplicity.

Although the regression process finds the best-fitting set of coefficients for a given set of control points, this fit may still not be very close. Inaccuracies in control point locations can produce large discrepancies between predicted and actual target coordinates. In this case, using the calculated transformation coefficients will produce registration errors. If discrepancies between predicted and actual target value are large, the correct procedure is to discard the most inaccurate (most deviant) control point and then recalculate the regression.

You can specify an accuracy threshold for your transformation functions. This threshold is specified in terms of a maximum *root mean square error* (RMSE). The RMSE for a given set of equations is computed by squaring all the distances between actual and predicted target coordinates, summing those squared distances, dividing by the number of points, and then taking the square root. Thus, the RMSE provides a sort of average discrepancy measure for the set of control points as a whole.

The default RMSE threshold in **Geography**⇒**Geometry**⇒**Calculate** is 0.500. This is a fairly common threshold. To require a more accurate set of transformation equations, specify a smaller value for the RMSE threshold. To be more lenient, specify a larger value.

Dragon computes a set of initial transformation coefficients using the full set of control points. Then it calculates the RMSE for the current transformation and compares that error measure to the threshold. If the error is higher than the threshold, the point that has the largest deviation between predicted and actual values is discarded, and the regression is recomputed using the reduced set of control points.

This cycle continues until either: 1) the error measure becomes less than or equal to the threshold or 2) fewer than four points are left. Four points is the absolute minimum that can be used to calculate an affine transformation. ·If the set of control points is reduced below four without reaching your error threshold, **Geography**⇒**Geometry**⇒**Calculate** gives up and does not produce a coefficient file. In this case, you have two choices: return to **GCP** and select new, hopefully more accurate, control points, or rerun **Calculate** specifying a more lenient error criterion.

If **Calculate** succeeds in finding a transformation that meets your error criterion, it writes the final set of coefficients to a coefficient file with the name you provide. This coefficient file will be used in the **Geography**⇒**Geometry**⇒**Register** operation to actually perform the geometric correction or registration.

Calculate displays a table summarizing its final results. However, the detailed results (coefficients at each stage, RMSE calculated, points discarded) are captured in a regression results report file. This file has the same name as the coefficient file you specify, but with the extension **.OUT**. It is a simple text file, which you can type or print at your leisure.

The *Output distance Unit* field, if non-blank, is stored in the coefficient file. **Geography**⇒**Geometry**⇒**Register** does not use this parameter, but does copy its value into the header of the registered image.

Register an Image (REG)

Choose **Register** to transform one or more **Dragon** image from one coordinate system to another. The second, reference coordinate system may correspond to a base map; in this case, the transformation is usually called *geometric correction*. Alternatively, the reference coordinate system may correspond to a second image. The transformation process in this case is often called *registration*.

There are a variety of situations where you might wish to register or geometrically-correct an image. If you plan to use the image data for updating a map or for comparison with a map, the image coordinate system needs to be congruent with the map. If you plan to import the image into a raster-based Geographic Information System, you will want to transform the image to the same dimensions and scale as the other layers in the GIS. If you need to compare two images from different sources within **Dragon**, for instance to perform a multi-temporal change analysis, you will normally need to register one of the images to the other.

In this third case, unlike the other situations described above, the choice of which image should provide the reference coordinate system may be arbitrary.

Geography⇒Geometry⇒Register transforms images to a new coordinate system using parameters included in a coefficient file. Normally, this file will be created by the **Geography⇒Geometry⇒Calculate** operation. It holds a set of six coefficients that summarize the mathematical relationship between the target image and the reference coordinate source (image or map).

The six coefficients (usually designated as **a0**, **a1**, **a2**, **b0**, **b1**, **b2**) in the parameter file implicitly define two equations, as follows:

```
Xt = a0 + (a1 * Xr) + (a2 * Yr)
Yt = b0 + (b1 * Xr) + (b2 * Yr)
```

These equations are the same ones described above, in the discussion of the **Geography⇒Geometry⇒Calculate** operation.

Xt and **Yt** are coordinates in the target image coordinate system. **Xr** and **Yr** are coordinates in the reference coordinate system. Using these equations, **Geography⇒Geometry⇒Register** can determine which point in the target image corresponds to a particular pixel in the output, transformed image.

Note that the calculated coordinates **Xt** and **Yt** will usually NOT be integers. They may fall entirely outside the bounds of the target (input) image. In this case, the resulting output pixel will be assigned a value of 0 (indicating no data was present). If the point **(Xt,Yt)** falls within the target image, but does not designate a unique pixel (because the calculated coordinates have fractional parts), a question arises concerning the image data value that should be assigned to the corresponding output pixel.

Geography⇒Geometry⇒Register provides two different methods for assigning values to the output pixels. The default method is known as *nearest-neighbor* assignment. **Register** selects the value from the input image pixel whose coordinates are closest to the calculated pixel coordinates, and assigns that to the corresponding output pixel.

The alternative method is known as *bilinear interpolation*. This method combines the data values from the four pixels surrounding the calculated pixel coordinates, using a weighted averaging technique that counts the closer pixel values more heavily.

The nearest neighbor method is faster than bilinear interpolation. It also more closely preserves the data value distribution of the input image. However, bilinear interpolation sometimes produces a smoother, more easily-interpreted image.

If the target image you are registering is a classified image, you should always use the nearest neighbor method. Bilinear interpolation, because it involves averaging, could produce class values in your output image that were not present in your input image.

In **Geography⇒Geometry⇒Register**, the *Method* radio buttons allow you to choose which method to use for value assignment.

Register also provides a set of parameters that are primarily useful when you are registering an image to a map. These parameters allow you to define the area to be registered, a scaling factor to be applied, and the units in which the registered image coordinates should be expressed.

The *Registration Area* fields allow you to specify X and Y coordinates for the upper left and lower right corners of the registered image, in reference coordinates. The *Output Pixel Size* fields allow you to specify the size of each pixel in the output image, in reference coordinates. Together, these two sets of parameters will determine the size and dimensions of the output image. The number of lines and pixels in the output file will be determined by taking the difference between the upper left and lower right coordinates, and dividing by the pixel size for the appropriate direction (X or Y). If you are registering an image to a map, you will usually want to specify both sets of parameters.

If you do not specify any values for these fields, then **Dragon** will generate an output image that includes the point with reference coordinates 0,0, and represents all the input data. This is probably what you will want, if you are registering one image to another; if you are correcting an image to a map, though, you should set the registration area to the coordinates of the corners of the map area you want the registered image to cover, and set the pixel size to the nominal scale of the input image. For instance, if you are registering an Enhanced Thematic Mapper image, the pixel size parameters should be set to 25 meters in each direction.

The ground control point files that you create by picking points on images in **Geography⇒Geometry⇒Ground Control Points** will always be in image coordinates, that is, in terms of columns (X) and rows (Y) in the image. If your reference image is georeferenced, you can transform the image coordinates to geographic coordinates using the **scaleGcp** utility which is included in the Tools directory. See the HTML documentation file for further details. If your GCP values are in geographic coordinates, you should specify values for the *Registration Area* fields, in order to get sensible and correct results.

Sometimes a reference coordinate system will have the a reversed directionality relative to the image coordinate system. For example, UTM northing coordinates are larger for points further north. This is the opposite sense from the image coordinate system used by **Dragon** (and most other image processing systems), where the Y coordinates of pixels further north (i.e. nearer to the top of the image) are smaller than those further south. In such a case, the Y coordinate entered for the upper left corner should be larger than the Y coordinate for the lower right.

To use **Geography⇒Geometry⇒Register**, you must specify the names of the image files that you want to register and the name of the coefficient file. You must also enter a prefix which will be used to name the output file or files. If you are registering a single image, the output file name will consist of the prefix, followed by **-R**. If you register multiple images at the same time (normally multiple bands of the same image), the output files will be named **prefix-R1.img**, **prefix-R2.img**, and so on.

You can optionally choose to use bilinear interpolation instead of the default nearest neighbor method, and you can specify a subarea to register if this meets your needs. Once you have entered the necessary information and clicked on **Ok** or pressed **<F1>**, **Dragon** will read the coefficient file and use it to transform the specified images.

Because the registration operation can produce more than one result image, it does not display any results, and does not change the data in image memory.

Geography⇒Geometry⇒Register automatically stores georeferencing information in the header of the output image. This information is derived from the coefficient file plus the **Register** input parameters. You can view the georeferencing information in the **Utility⇒List** operation, and change it with the **Utility⇒Header** operation.

6.6   Polygon Operations

Choose **Geography⇒Polygons** to draw polygons or polylines on an image and use these vector features for various purposes.

Vector Overlay Creation (VEC)

Choose **Geography⇒Polygons⇒Vector** to create images that consist of linear features, or to add linear features to an existing image. The linear features presently implemented include multi-segment lines (or *polylines*), polygons and filled polygons, and, in **Dragon Professional**, point features (or *markers*). **Vector** allows you to specify a background image to serve as a guide. This background can assist you in tracing out road networks, rivers, water bodies, and other spatially distinct features. You can choose the color you want to use for drawing; this color corresponds to a data value that will be assigned to all points on the line or inside the filled polygon in the final image. Finally, you can specify that the vector features you create should be merged into the background image, producing a hybrid image that includes both sensor-based image data and user-created vector features.

Geography⇒Polygons⇒Vector can serve several purposes. It allows you to highlight or delimit features of interest on an image. It allows you to create complex mask images that can be used in **Enhance⇒Mask** to select areas of interest and screen out irrelevant areas. Finally, **Vector** can be used to create data layers for a GIS system, based on information in a remotely sensed image or in an imported vector file. For instance, **Vector** could be used to delimit field boundaries for an agricultural study, or settlement areas for an urban planning database.

The results of **Vector** are normally converted to raster form, as an image. You can also choose to save the linear feature information in vector form, as a **Dragon** vector file. In addition, you can use **Vector** to add new vector features to an existing vector file. The combined vectors can then be saved as a new vector file and/or rasterized.

Like most operations, **Vector** leaves its results in image memory formatted as a standard image. You should save them as an image file (using **File⇒Save**) for overlay displays or to export them to a GIS package.

The images created by **Vector** have a filetype of *L*, for *Layer* files. This signals that data in these files should not be stretched, since individual data values may represent attribute codes rather than continuously-varying physical measurements (as in a remotely-sensed image file). Layer (*L*) files are handled somewhat differently from classified (*C*) files in some operations, since a layer file can have pixels of any value, while values in a classified file are limited to the maximum possible class number.

Vector Parameters

The parameters in the **Vector** response panel allow you to control details of **Vector** behavior. First, you can specify a background image (the *reference image*) to guide your vector creation. You can select =**M**, =**R**, =**G**, =**B**, or =**C** as the background image. (You will not be able to merge vectors into the =**C** image.)

You are not required to enter a file for this parameter; if you do not, the background will be black. Also, if you do not specify a background image, you can specify the size of the blank image in which you will define vectors. The default size is 1024 x 1024. If you are running **Dragon Professional**, you can specify any size up to 8000 x 8000.

After you finish creating points, lines and polygons with **Vector**, these linear features are processed into standard image format. Normally, this new image will become the new current image (=**M**), replacing the image you have been using as a background. Alternatively, you can choose to merge the vector features with your reference image. Note that if you do check the box for the the *Combine..?* question, some data in the background image will be overwritten by vector data.

Note
You can effectively edit pixel values in the background image by combining vector features with the background image.

The parameters available to you vary depending on whether you check the *Combine...* box. If you leave this box unchecked (the default), you have the option of entering two different color files. The *Color File for reference image* parameter allows you to control the display of the background image, whereas the *Color File for Vectors* item determines how attribute values will be mapped to colors when drawing. As long as you are not merging the vectors into the background, you can control these two sets of colors independently. If your background image is a single band of image data, you may wish to display it in shades of gray using **defgra.clf**.

If you check the box, indicating that you want to merge the vectors into the background, you can only select a single color scheme, which will be used for both background and vectors. This is to insure that the final, merged image will look the same as the image during the process of defining the vector features. Note that the file type of the resulting image ('Image', 'Classified' or 'Layer') will be determined by the initial file type of the background image.

In most operations, the image size is determined by the image being displayed, so the user need not be aware that a choice is being made. In **Vector**, however, you may be creating an entirely new image. If you specify a background image, the dimensions of the output image will be the same as the background image dimensions. If you do not specify a background image, you can specify the dimensions for your output image. The default is 1024 lines by 1024 pixels.

The *Apply Vector File* parameter allows you to specify a source for an initial set of vectors to be displayed on the image before interactive definition of linear features begins. These vectors will be rasterized into the final image along with any new features you create.

The *Vector File to Create* item, if specified, indicates that you want to save the linear features defined in **Vector** in vector as well as raster form. If you also specify a value for the *Apply Vector File*, these original vectors will be copied into the output vector file.

Geography⇒Polygons⇒Vector Interaction

Vector is an interactive operation. After you provide the desired parameter values, **Dragon** will display the image you specified for the background (if any). If you specified a file name for the *Apply Vector File* field, the vectors in this file are read and displayed.

Dragon displays a button panel very similar to that available for training, which includes the following buttons:

- **Rasterize** (**<F1>**): Completes the interaction and initiates rasterization, and optional saving of vector files.
- **Select Color** (**<F3>**): Displays a dialog where you can type in an attribute number to use for drawing. The color associated with the attribute is displayed in a sample box next to the data entry field.
- **Help** (**<F4>**): Displays general instructions for defining vectors.
- **Accept Point/End Line** (**<F6>**): If only one point has been selected, accepts that point as a marker feature. If more than one point has been selected, ends the current polyline at the last selected point. (The label on this button changes depending on the current state of point selection.) (Only **Dragon Professional** supports the creation of point features.)
- **Close Polygon** (**<F7>**): Draws a final line segment from the last selected point in the current polyline to the first, creating a closed polygon.
- **Fill Polygon** (**<F8>**): Fills the last created polygon with the currently selected attribute/color.
- **Delete** (**<F9>**): Deletes the last created figure.
- **Cancel** (**<F10>**): Exits from the **Geography⇒Vector** interaction without rasterizing image or saving vectors in vector file.
- **Region** (**<F11>**): Select a region for display at full resolution (1:1) so that points can be picked precisely (in **Dragon Professional** only).

Buttons that are not relevant are disabled. For example, the **Fill Polygon** button will not be enabled until you select **Close Polygon**.

It may take a few moments for the button panel to show up. Note that you cannot begin to define vector features until it is visible.

To start a feature, move the cursor to the desired point on the image and select that point (by pressing the **<Ins>** key or the left mouse button). **Dragon** will display a rubberbanding line that starts at the selected point and follows the cursor as you move. This line is always a color which contrasts with the background. Position the cursor at the next point and select it. A line segment will be drawn from the first point to the second point. Each time you choose a new point, **Dragon** draws a new segment in the selected color.

When you have selected all the desired points, you must signal that you are finished. Press **<F6>** or click on the **End Line** button to end a line at the last selected point. Press **<F7>** or click on the **Close Polygon** button to create a polygon by drawing a line from the last selected point to the first selected point.

If you are using **Dragon Professional** and you want to create a single point feature, click on **Accept Point** or press **<F6>** immediately after choosing the first point. An "X" will appear, whose center marks the point that you selected.

After you finish a line or polygon, you have several possible options. You can begin a new linear feature by positioning the cursor and selecting a starting point. If you have just completed a polygon, you can fill it with a solid color by clicking on **Fill Polygon** or pressing **<F8>**. You can also delete the last feature you created by clicking on **Delete** or pressing **<F9>**. When you press **<F9>**, **Dragon** will redraw the image and then redisplay all the features you have created so far, except for the last one created.

Press **<F3>** or click on the **Select Color** button to choose a new color and attribute value for drawing. **Dragon** will display a dialog where you can enter a number between 0 and 255. As you type into the data entry field, the color corresponding to the value you have entered will be displayed in the sample color box to the right of the field.

The color that corresponds to a particular value is determined by the color scheme that you specify for *Color File for Vectors*. The default is the **def3ba** color scheme. The default, starting value is 211, which is a bright yellow in the **def3ba** scheme.

You cannot change colors in the middle of drawing a line; the **Select Color** button is disabled. However, you can draw the boundaries of a polygon in one color and then fill it with another by selecting the fill color after completing the polygon with **Close Polygon** or the **<F7>**, then click on **Fill Polygon** or press **<F8>**.

If you are using **Dragon Professional** you can choose a full resolution (1:1) region at any time except while you are actually drawing a feature. Click on **Region** or press **<F11>**, then use the mouse to select a region of the image. You will not be able to select a region larger than will fit on your screen at the 1:1 resolution. To return to an overview of the entire image, press the *right* mousebutton while your mouse is anywhere within the image. See *Chapter 2* for more details.

As is the case with other interactive operations, you can zoom the image in **Vector**. This can make it easier to select specific points on the image. Use the **Zoom** menu on the viewport to zoom in or out.

After you have created all the desired linear features, finish your **Vector** session by clicking on **Rasterize** (or **<F1>**). There will be a slight delay while **Dragon** *rasterizes* the linear features you have created (that is, writes them out to image memory in the row/column format used for images). Then **Dragon** will calculate statistics for this rasterized image. Note that if you display the histogram for the image in memory after executing **Vector**, you will most likely see a large peak at 0, with smaller peaks at each attribute value you used for drawing. The only time you will not see this kind of pattern is if you chose to merge your vectors with a background image.

If you specified a filename for the *Vector file to create* field, clicking on **Rasterize** will write all the vector features that you created to the specified vector file. If you entered a filename for *Apply Vector File*, vector features from that input file will be copied to the output vector file as well.

Finally, **Dragon** will display the rasterized image. If you chose to merge the features with the background, this image should appear identical to the display while you were drawing. If you did not choose to combine the vectors with the background, you will see your vectors displayed on a field of black (unless you chose a color scheme where the value 0 is assigned to some color other than black).

If you select **Cancel**, you will cancel all of the **Vector** operations. No rasterized image or vector file will be created. The background image, if any, will be redisplayed.

Vector can be used with both *georeferenced* and non-georeferenced images. A georeferenced image has information in its header that specifies the relationship between image coordinates (line and pixel) and geographic coordinates (e.g. UTM). If you use a georeferenced image as a background in **Vector**, any vector file that you create will hold geographic coordinates. If the background is not georeferenced, a newly created vector file will hold image coordinates.

When you specify a filename for *Apply Vector File*, **Dragon** checks to make sure that the features in that file are geographically congruent with the background image file (if any). If both the vector file and the image file are georeferenced, but their geographic extents do not overlap (that is, they refer to completely different locations on the earth), **Dragon** will display an error message and exit from **Vector**.

Measure Lengths and Areas (MEA)

Choose **Geography⇒Polygons⇒Measure** to measure various attributes of the image. At present, you can display the lengths of (poly)lines, the perimeters and areas of polygons, a *radiometric profile* along a line or polygon, and the histogram of data within a polygon. By default, **Measure** displays the measured

Copyright ©Global Software Institute

values in terms of pixels. However, you can supply conversion factors appropriate for your image so that **Measure** will report image measurements in a unit of your choice (e.g. meters, miles). If your background image is georeferenced, the georeferencing metadata will be used to supply a default conversion factor and units.

Measure is useful for analyzing classified images. It allows you to select specific areas that belong to particular classes and calculate their areas, independent of other pixels belonging to that class. **Measure** can also provide rough measurements for engineering studies, such as measurements along a road, a river course or a ridgeline.

Using **Geography⇒Polygons⇒Measure**, you can also display a *line profile*. A line profile (sometimes called a *radiometric profile* or a *transect*) is a graphical representation of the variation in image data values over space. Using the line profile feature, you can see the pattern of image data values along a linear path (which can consist of multiple line segments). If your 'image' data happen to be digital elevation data, the line profile can show you the shape of the landforms along the selected line.

Measure can also display the histogram of a polygon region. The histogram shows the relative number of pixels within the region having each data value.

You can capture the line profile and area histogram data in an ASCII text file, in a format designed for easy import into spreadsheets, statistics packages or graphics packages. To use this capability, simply specify the name of an output data file in the **Measure** response panel (*File for data Output* field). If you enter a file name, data from all profiles and histograms in the current **Measure** session will be written to that file. (For information on the format of **Measure** data files, see the *File Formats* chapter of the *User Manual*.)

The parameters required by **Measure** are relatively simple. You must supply the name of an image file that will serve as your background for tracing out linear features and polygons.

Note that you can enter the special filename **=C** in **Measure**, to perform measurement against a composite image background. However, the line profile and area histogram functions are disabled when the background is a composite image, since these operations currently work on a single image at a time and **=C** really refers to three images at once.

You can provide the name of a distance unit. If you provide a unit, you should also provide scaling factors. The scaling factors are real (i.e. decimal, or floating-point) numbers that express the number of units per pixel in the X and Y directions. For instance, if you plan to measure areas on an image derived from Thematic Mapper data, you might enter **Meters** for your desired unit. In this case, you would want to enter 30.0 for the X and the Y scaling factor parameters, since each pixel in a TM image is normally 30 meters square. If, on the other hand, you specified a unit of Kilometers for this image, you would enter 0.03 as your scaling factors (since each 30 meter pixel represents 0.03 kilometers).

Finally, as discussed above, you can specify a filename for an output file where **Dragon** will store profile and histogram data values.

Geography⇒Polygons⇒Measure is very similar to **Geography⇒Polygons⇒Vector**, but a bit simpler. First, **Dragon** displays the background image.

Dragon then displays a dialog which has a text area for displaying measurements, and a button panel which includes the following buttons:

- **Finished** (**<F1>**): Completes the interaction and returns to the menuing system.
- **Help** (**<F4>**): Displays general instructions for defining vectors.
- **End Line** (**<F6>**): Ends the current polyline at the last selected point.
- **Close Polygon** (**<F7>**): Draws a final line segment from the last selected point in the current polyline to the first, creating a closed polygon.
- **Measure** (**<F8>**): If the last created figure was a polygon, measures the perimeter and area of the last created polygon and displays them in the text area. If the last figure was an unclosed polyline, calculates and displays the length.
- **Refresh/Delete** (**<F9>**): Deletes all previously drawn features. During the creation of a new linear feature or polygon, deletes the feature in the process of being drawn.
- **Region** (**<F11>**): In **Dragon Professional** only, enables selection of a region for display at full resolution (1:1) so that points can be picked precisely.

Buttons that are not relevant are disabled. For example, the **Measure** button will not be enabled until you select **Close Polygon** or **End Line**.

It may take a few minutes for the dialog with the button panel to show up. Note that you cannot begin to choose points until it is visible.

To select a point, move the cursor to the desired position and press the left mouse button or the **<Ins>** key. A rubberband line is anchored to the selected point and follows the cursor until you select the next point. To finish a linear feature, click on **End Line** or press **<F6>**. To complete a polygon by drawing a line from the last selected point to the first, click on **Close Polygon** or press **<F7>**.

Each figure's starting point is labeled with a number, so that you can correlate the measurement information in the text area to the figures on the image.

Geography⇒**Polygons**⇒**Measure** differs from **Geography**⇒**Polygons**⇒**Vector** in the options it provides after you complete a line or polygon. If you click on **Measure** or press **<F8>** after completing a figure, **Dragon** measures that figure and displays the measurement data as the top line in the text area. If the last figure drawn was a series of line segments, **Dragon** displays the length. If the last figure was a polygon, **Dragon** displays the perimeter and the area. As discussed earlier, **Dragon** uses the units you request and the scaling factors you provide for reporting these measurements. Area measures are reported in *square units*. For instance, if you specified *miles* as your unit, **Dragon** would report area in *square miles*.

To display a line profile that shows the data value variation along the lines of the last feature, click on click on **Profile** or press **<F7>**. The line profile will be displayed in a separate window.

To display a region histogram showing the distribution of values within a polygon, define a polygon, close it by clicking **Close Polygon** or the **<F7>**, then click on **Histogram** or press **<F6>**. The histogram will be displayed in a separate window, with a title that identifies it by feature number.

You can repeat the feature drawing and measurement sequence as many times as you wish. Note that you can only measure the last figure created. However, the previous measurements remain in the text area for your review. You can measure a feature and then graph its line profile or area histogram, or vice versa.

If you are using **Dragon Professional**, you can choose a full resolution (1:1) region at any time except while you are actually drawing a feature. Click on **Region** or press **<F11>**, then use the mouse to select a region of the image. You will not be able to select a region larger than will fit on your screen at the 1:1 resolution. To return to an overview of the entire image, press the *right* mouse button while your mouse is anywhere within the image. See *Chapter 2* for more details.

After you have been using **Geography**⇒**Polygons**⇒**Measure** for several sets of measurement, your image may begin to appear rather cluttered with various lines and polygons. You can click on **Delete** or press **<F9>** to redraw a clean copy of the image.

When you are finished with **Measure**, click on **Finished** or press **<F1>**. **Dragon** will return immediately to the menuing system.

Note that like the other Geography operations, **Measure** allows you to zoom the background images. This can make it easier to select specific pixels. Use the viewport **Zoom** menu to zoom in and to restore the image.

Cookie Cutter Subset Extraction (COO)

Choose **Geography**⇒**Polygons**⇒**Cookie Cutter** to extract an irregularly-shaped subset from an image. In **Cookie Cutter**, you draw a polygon containing all the area you want to retain in your cookie-cutter subset. When you indicate that you are happy with this area selection, **Dragon** sets all the pixels outside this polygon to a background value (which you can select). If your polygon is considerably smaller than the original image, **Dragon** can 'clip' the resulting image to the smallest rectangle that contains the polygonal subset. This will result in an image file that requires less disk space.

The main reason for using **Cookie Cutter** is to define a 'study area', eliminating areas that are not of interest for your application. For example, if you are working on a county-wide wetlands assessment study, you might want to exclude all areas outside the county boundaries. If your primary interest is in coastal sedimentation, you might use **Cookie Cutter** to exclude all land area from a coastal image, leaving you with an ocean-only image.

Cookie Cutter can only work on one image at a time. (Thus, you cannot specify =**C** as the file to use for background.) However, once you have created a cookie-cutter image. you can save that image and then use it as a mask (using the Enhancement operation **Enhance**⇒**Mask**), to exclude the same areas from any number of other image bands. (Note: for this strategy to work, you must set the background value to zero (the default) and NOT clip to the bounding rectangle). An alternative strategy is to create a mask image in **Geography**⇒**Polygons**⇒**Vector**, then use **Mask** to mass-produce your cookies.

Geography⇒**Polygons**⇒**Cookie Cutter** is similar to **Geography**⇒**Polygons**⇒**Vector** and **Geography**⇒**Polygons**⇒**Measure**, but considerably simpler, since in **Cookie Cutter** you can define only a single polygon. **Dragon** displays the image you select for subsetting Then **Dragon** displays a button panel very similar to that available for vector definition and measurement, which includes the following buttons:

- **Accept** (<**F1**>): Completes the interaction and initiates the cookie-cutting of the image.
- **Help** (<**F4**>): Displays general instructions for defining the cookie polygon.
- **Close Polygon** (<**F7**>): Draws a final line segment from the last selected point in the polyline to the first, creating a closed polygon.
- **Delete** (<**F9**>): Deletes the current figure, redisplaying a clean background image.
- **Cancel** (<**F10**>): Exits from the **Cookie Cutter** interaction without executing the subsetting operation.
- **Region** (<**F11**>): Select a region for display at full resolution (1:1) so that points can be picked precisely (**Dragon Professional** only).

Buttons that are not relevant are disabled. For example, the **Accept** button will not be enabled until you select **Close Polygon**.

It may take a few minutes for the button panel to show up. Note that you cannot begin to define the cookie polygon until the panel is visible.

To define your irregular subset boundary, move the cursor to the first vertex in the cookie-cutter polygon and select it (by pressing the left mouse button or <**Ins**>). A rubberband line appears, anchored to the selected point, and follows the cursor as you move it to the next desired vertex. Continue selecting points until you have outlined the desired shape in the desired location, then click on **Close Polygon** or press <**F7**> to close the polygon.

After you close the polygon, you have two choices. Click on **Accept** (or press <**F1**>) to accept the polygon and continue with the subset extraction. Click on **Delete** (or press <**F9**>) to reject the polygon, in which case **Dragon** will redisplay the background image without the lines and you can start again.

If you are using **Dragon Professional**, you can choose a full resolution (1:1) region at any time except while you are actually drawing a feature. Click on **Region** or press <**F11**>, then use the mouse to select a region of the image. You will not be able to select a region larger than will fit on your screen at the 1:1 resolution. To return to an overview of the entire image, press the *right* mouse button while your mouse is anywhere within the image. See *Chapter 2* for more details.

When you accept the polygon, **Dragon** pauses a few moments to do calculations. Then it erases the original image and displays the results of the subsetting. Areas inside the cookie-cutter polygon will remain unchanged. Areas outside will be set to the background value you selected. The color that they appear will depend upon the color file used to display the original image. A background value of zero will almost always appear black.

If you requested that **Dragon** clip the image to the smallest bounding rectangle, that rectangle will be displayed at the upper left corner of the screen. This may appear as if the subsetted area has moved upward or leftward. If you save the cookie cutter results and then examine the image statistics with **Utility**⇒**List**, you will see that the cookie-cutter image has fewer lines and/or pixels per line than the original.

Note that you cannot create a subset from the area outside the polygon in **Geography**⇒**Polygons**⇒**Cookie Cutter** (from the 'dough' rather than the 'cookie'). However, you can achieve much the same effect in **Geography⇒Polygons⇒Vector**, by checking the *Combine Vectors with Background?* box, drawing your polygon, and then filling it with color number 0.

Like other interactive geography operations, **Cookie Cutter** allows you to zoom the image before defining your polygon.

Cookie Cutter places the subsetted image in the main memory image (=M). After completing the "cookie-cutter" process, you can display or save this image, just as you can the results of any image processing operation.

6.7 Raster Modeling Operations

Choose **Geography⇒Raster Modeling** to create or combine layers representing thematic as well as radiometric data.

Slope Calculation (SLO)

Choose **Geography⇒Raster⇒Slope** to create a slope image from an image file that contains elevation data. *Slope* is a measure of the steepness of the land surface at a particular point on the earth. A point located in a flat area will have zero slope.

The slope at a particular location is an extremely important parameter for many kinds of geographic decision making. For instance, one would not want to site a building or practice agriculture in a location with a steep slope. Slope is also correlated with vegetation type and thus may determine animal habitat. Calculating a slope image is often a first step in series of raster modeling operations.

Point number 1: X = 59, Y = 99 (489165.72, 2080698.06)		
chiangmaidem		
1513	1504	1491
1507	**1497**	1487
1509	1491	1480

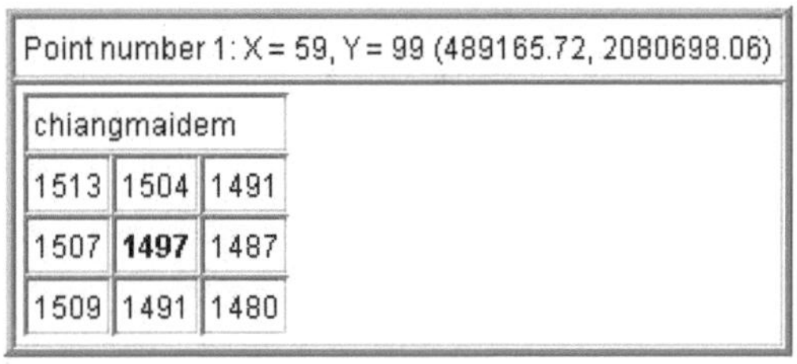

Figure 6.3: *Elevation values in a pixel neighborhood*

Dragon determines the slope at a particular pixel position by calculating the difference in elevation values between all pairs of neighbors that are on opposite sides of the pixel. The largest (absolute) difference will be used as the slope for that pixel.

As an example, consider Figure 6.3. This shows the elevation values in the neighborhood around the pixel at located at line 99, pixel 59. **Dragon** will compute the difference in values for pixels that are on either side of the center pixel, horizontally (1507 - 1487 = 20), vertically (1504 - 1481 = 13), and along both diagonals (1513 - 1480 = 33 and 1509 - 1491 = 18). Clearly, in this example, the northwest to southeast diagonal has the largest difference. Therefore this becomes the value for this cell in the output image.

Dragon scales all the calculated slope values to be in the range of 0 to 255.

To calculate a slope image, you must specify the name of the input image file. This image file should be a Digital Elevation Model (DEM) or Digital Terrain Model (DTM). That is, the values in each cell should represent height measurements, not measurements of reflected electromagnetic radiation. If you apply the Slope operation to a regular remotely sensed image, the results will not make any sense.

The Slope operation displays the calculated slope image. You can choose whether to display the image in gray or color, the color scheme that should be used, and the viewport in which the result image should be displayed.

Aspect Calculation (ASP)

Choose **Geography⇒Raster⇒Aspect** to create an aspect image from an image file that contains elevation data. *Aspect* measures the direction that a slope faces at a particular point on the earth. A point located in a flat area will have an undefined aspect, but otherwise, aspect will be expressed as a direction such as 'North', 'North East', and so on.

The aspect at a particular location is an important parameter for some problems. For instance, solar panels for generating electricity should be located on slopes with an aspect that receives the maximum amount of sunlight. The aspect at a point will determine what can be seen from that point. Thus, aspect may be a consideration in architectural design.

To determine the aspect at a particular pixel position, **Dragon** first determines the slope. As explained in the previous section, this calculation identifies which neighbors of the focus pixel have the largest difference in elevation values. Unlike the slope calculation, aspect calculations also need to consider the direction of the difference. In the example presented in the previous section the northwest to southeast diagonal has the largest slope value. Since the northwest pixel is larger than the southeast pixel, we can determine that the slope faces southeast. If the direction of the difference were reversed, then the slope would face northwest.

Dragon uses a set of standard class numbers to represent aspect values. An aspect value of zero means that the pixel is on a flat surface, has zero slope and thus has an undefined aspect. A value of one indicates a north-facing slope. A value of two indicates a northeast-facing slope. Three means the slope is east-facing. **Dragon** stores the aspect directions as class names in the output image, so you can display a legend that identifies the various aspect values.

To calculate an aspect image, you must specify the name of the input image file. This image file should be a Digital Elevation Model (DEM) or Digital Terrain Model (DTM). That is, the values in each cell should represent height measurements, not measurements of reflected electromagnetic radiation. If you apply the Aspect operation to a regular remotely sensed image, the results will not make any sense.

Because the results that it calculates are in the form of a classified image, **Dragon** always displays them in color. By default the display uses the color file **defaspct.clf**. However, you can choose a different color scheme if you wish.

Buffer Calculation (BUF)

Choose **Geography**⇒**Raster**⇒**Buffer** to create an image that identifies a buffer around some target locations. A *buffer* is a region in which all points are less than or equal to a specified distance (the buffer distance) from some target locations. For instance, you might want to a create a buffer that represents all locations within 30 meters from a road.

Buffers are widely used in geographic modeling. In some situations, a buffer identifies an area of potential interest. You might want to identify the area that could be served by a hospital by calculating a buffer that represents all locations within two kilometers of the hospital. In other situations, a buffer represents a zone of exclusion. In looking for a site for a new landfill to handle a town's waste, you might want to make sure that all potential locations are more than 500 meters from any water body. In this case, the buffer identifies locations that should be excluded from consideration.

To calculate a buffer using **Dragon**, you must specify the name of an input image file. You also need to identify which pixels are target pixels, that is, starting points for calculating the buffer. You identify target pixels by specifying a *Target value*, which is some value between 0 and 255. All pixels that have the specified value in the input image file will be used as targets.

Usually, the input image used for buffering will be a classified image or a layer file. In these types of files, the image data values represent classes or categories. For instance, in a classified image, the water class might be represented as class value 3. To create a buffer around all water, you would specify 3 for the *Target value* parameter. As another example, you might use the **Geography**⇒**Polygons**⇒**Vector** operation to trace out all the roads in an image. Then the resulting rasterized layer could serve as a starting point for creating buffers around roads.

You also must specify the size of the buffer, that is, the *Buffer distance*. The buffer distance should be expressed in geographic units, such as meters or feet. **Dragon** will assume that the buffer distance you enter refers to the units stored in the georeferencing section of the image header file. If the input image is not georeferenced, the buffer distance will represent pixels.

The maximum distance that **Dragon** can represent is 254 pixels (or the equivalent in appropriate geographic units). If you specify a larger buffer distance, **Dragon** will stop calculating when it reaches its limit.

The Buffer operation allows you to choose one of three different distance metrics: **Euclidean distance**, **Neighborhood distance** or **City block distance**. Euclidean distance provides the most accurate results, but also takes the longest to calculate. The distance metric that you choose will affect the shape of the result buffer. Euclidean distance buffers look rounded at their edges, neighborhood buffers have a square shape, and city block buffers have a diamond shape.

The Buffer operation displays its results as an image. In the result image, target pixels have the value 255. Buffer pixels have the value 127. All other pixels will be set to zero. You can control the display in the usual ways: choosing gray or color, selecting a color scheme, and choosing the viewport. By default, the operation displays its results in color, using a standard color file called **defbufr.clf**.

Combine Layers (COMB)

Choose **Geography⇒Raster⇒Combine Layers** to create an image whose values depend on some logical relationship between corresponding values in the input layers. **Combine Layers** applies a set of user defined rules to the input layers in order to calculate the output values. These rules specify an output value to be applied when each input value falls within a specific range of values. An output value may depend on all the input layers, or only a subset of them. All cells which do not match any rule are set to zero.

For example, suppose that we have two classified images of the same area to use as input. One identifies water in a rainy season image as class 1. The other identifies water in a dry season image as class 14. Suppose we want our output image to identify pixels that are water in the rainy season but not in the dry season. We can accomplish this using two range-based rules:

- Assign 255 if $(1 <= Image1 <= 1)$ and $(0 <= Image2 <= 13)$
- Assign 255 if $(1 <= Image1 <= 1)$ and $(15 <= Image2 <= 255)$

In other words, we will assign a value of 255, indicating this is a pixel of interest, to all the pixels whose value is 1 in the first input image and whose value is **not** 14 in the second input image.

Figures 6.4 through 6.6 show the input images for this example, and the results. In the first, rainy season image, the water is blue. In the second, dry season image, the water is yellow. The result image shows in white all pixels that were water in the rainy season but not water in the dry season. The shrinkage of the reservoir is obvious.

The **Geography⇒Raster⇒Combine Layers** operation is conceptually similar to the table-based recoding in **Classify⇒Recode**. In both cases, the recoding is conditional, based on input values. In the **Combine Layers** operation, however, as many as twelve input layers can be used to determine the values in the output layer.

The **Combine Layers** operation can be used for a variety of different types of modeling. It can be used to implement some *multi-criteria decision models*. These are models in which many different attributes are examined and weighed, usually in order to identify locations that are appropriate for some new type of use. For instance, we might be trying to locate areas that would be appropriate for a new state park. Our criteria might be that all areas in the park must be within half a kilometer of some river or lake, that the areas in the park must currently have a landcover type of either forest or grassland, and that the areas that will become park must currently be owned by either the state or individual municipalities. If we have raster layers that represent a buffer around the water features (which we could create with **Dragon**), landcover type, and ownership category, we could define rules to identify candidate park locations.

The **Combine Layers** operation can also be used to create a *decision tree classification*. Classical supervised classification methods such as maximum likelihood apply the same processing to all image pixels, and consider all possible classes for each one. A decision tree approach to classification can use different tests, strategies or rules for different classes. For instance, a decision tree designed to work with Thematic Mapper data might begin with a test that says if Band 5 is less than 10, the pixel should be classified as Clear Water. Then a second test could categorize pixels as Healthy Vegetation if their values were less than some threshold in Band 2 and greater than some threshold in Band 5. In **Dragon**, the rules used to combine layers are considered one by one, in order. The first rule that is satisfied determines the output value. Thus, with a bit of care, we can order the rules to act like a set of hierarchical tests.

Figure 6.4: *Input image 1: rainy season*

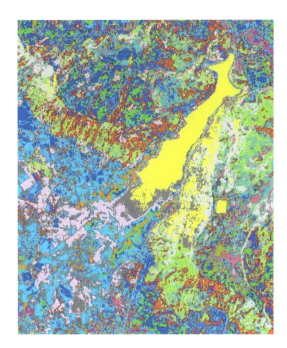

Figure 6.5: *Input image 2: dry season*

A third important application for **Combine Layers**, illustrated by the wet season/dry season example above, is *multitemporal change detection*. A common strategy in change detection is to create two registered, classified images that represent the same area during different time periods. Clearly pixels that have different values in the two classified images represent a change, but determining and describing the nature of the change may be difficult. With the **Combine Layers** operation it is straightforward to pick out pixels which, for example, changed from forest to cleared, or from open land to developed. Specific change vectors can be highlighted and other changes of less interest can be ignored.

In order to use **Combine Layers** you must have created a rule file. The **Geography⇒Raster⇒Rules** operation, described in the next section, can be used for this purpose. You must select the input images that you will combine. The multi-selection file chooser (see **Enhance⇒Principal Components Analysis** for details) can be used to type or choose the different images, and to reorder them if necessary. The order of your input images must match the logic of your rules, or you will not get the results that you expect.

In addition to the input images and the *Rule file*, you also have the opportunity to specify a *Default value*. This is the value that will be assigned to output pixels if the corresponding input pixel values do not match any of the rules in the rule file. The default value is 0, which corresponds to the **Dragon** convention for 'unclassified' pixels.

Geography⇒Raster⇒Combine Layers displays the resulting image, which is a Layer type image file. You can choose the desired color scheme and viewport.

Create Rules (RULE)

Choose **Geography⇒Raster⇒Rules** to create a set of rules to use in the conditional recoding operation, **Geography⇒Raster⇒Combine Layers**.

Rules are stored in **rule files**. A rule file is a text file which has one rule per line. Rules have a simple, human readable form:

```
newVal IF (b1min ≤ B1 ≤ b1max) & (b2min ≤ B2 ≤ b2max) & ...
```

97

Figure 6.6: *Result of rule-based combination*

In this format, 'newVal' is the value that will be assigned if the rule is satisfied. Each condition in parentheses specifies a set of limits for one of the bands. The input value for the corresponding rule must fall within the specified limits in order for the rule to be satisfied.

Not every band has to be mentioned in a rule. A band that does not have a condition will not be checked; any value in that band will allow the rule to be satisfied. A rule can specify that a band must have a single, specific value by making the minimum and maximum limits the same.

For example, here is very simple rule:

    14 IF (3 ≤ B1 ≤ 3) & (3 ≤ B3 ≤ 6)

This rule will assign an output value of 14 if the first input band value is 3, and the second band is between 3 and 6 inclusive.

The user interface for creating rules is quite different from most of the **Dragon** operations. As shown in Figure 6.7, the rule creation screen consists of two main areas. At the top are a set of fields where you can enter the limits for various bands. There is also a field for setting the output value. When you have entered limits for all the bands that are needed in a particular rule, click the **Add** button to add the rule to the list. The rule will show up in the text area below the limits fields. Meanwhile, the limits fields and the output value field will be cleared to their default values.

If you are satisfied with the rule, you can go ahead and specify another output value and set of limits. There is no limit to the number of rules that can exist in a rule file.

If you enter a rule that has mistakes, click on the rule to select it and then click the **Delete** button. The most common error is to forget to specify the output value. You can also change the order of the rules by selecting one of the rules and clicking **Move Up** or **Move Down** buttons until the rules are in the desired order.

When you are finished defining the rules, enter a name for your rule file, and click on the **OK** button. Your rules will be saved and you can use them immediately.

There is no user interface mechanism that allows you to edit existing rules. As noted above, if you make a mistake, you must delete the rule and then enter it again. However, rule files are simple text files, and they can be edited using normal text editor programs such as Emacs or Notepad. If you do edit a rule file, you

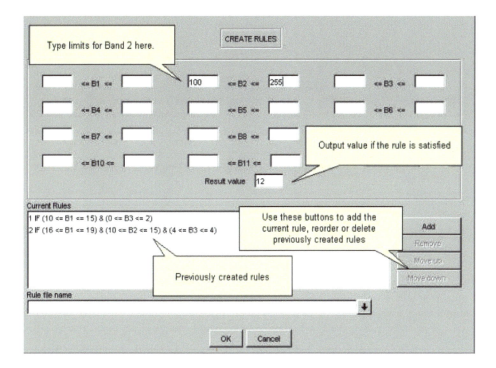

Figure 6.7: *User interface for creating rules*

must be careful to keep each rule on its own line. If you break a rule in the middle by inserting a carriage return, **Dragon** will display an error message when it tries to read the rule file.

6.8   Exercises

This section offers a set of hands-on exercises with some of the **Dragon** geography operations, in order to let you practice using these functions and see their results.

1. Start your **Dragon** system. Choose **Geography**⇒**Polygons**⇒**Vector**. Enter the following parameters:

   - Image File: amhphoto.img
   - Vector File to create: amherstRoads.vec

   Click the **OK** button to start the operation. **Dragon** displays the image, a scanned aerial photo of Amherst, Massachusetts, in gray. **Dragon** will also display the button panel to the left of the image. Click on the **Select Color** button or press **<F3>**. **Dragon** pops up a dialog where you can type a color number and see the result. Type **150**. The color box turns red. Click **OK** to dismiss the dialog.

   You will now create a road vector by tracing one of the roads in the image. Place the cursor near the top of the image, at the start of the long road that runs between the stadium (large white square building) and the racetrack (white oval). Click the mouse. **Dragon** starts a line at the point you clicked. Move the mouse down to the first intersection and click again. **Dragon** draws a line segment in red from the first point you picked to the second. Click on the next intersection down. Once again, **Dragon** draws a line segment. Continue until you get to the intersection of your road with the other road running at a forty five degree angle from south west to north east. Then click the **End Line** or press **<F6>**. The polyline ends at the last position of the mouse.

   Digitize (that is, draw) another linear feature following the course of the forty five degree line. Digitize the three main roads that are roughly parallel to the first road you drew. Continue until you have captured most of the main roads you can see.

Click the **Select Color** again. Change the color to **5**. The color box turns bright blue. Dismiss the dialog. Now begin to outline the stadium, clicking at each corner. After you have clicked on the fourth corner, click **Close Polygon** or press **<F7>**. **Dragon** draws the final segment to create a polygonal outline surrounding the stadium.

Click on **Fill Polygon** or press **<F8>**. **Dragon** fills the polygon you just drew with the current color. Click **Delete** or press **<F9>**. The filled polygon around the stadium disappears.

Click the **Rasterize** or press **<F1>**. The button panel disappears. After a few moments, **Dragon** displays the vector features you created, with a black background. Choose **File⇒Save** and save this image with the name **amherstRoads.img**.

2. Choose **Geography⇒Polygons⇒Vector** again, enter the following parameters, and click **OK** to start the operation.

    - Image File: amhphoto.img
    - Apply Vector File: amherstRoads.vec
    - Vector File to create: amherstRoads2.vec

Dragon displays the image again. Then it reads the vector information stored in **amherstRoads.vec** and displays those vectors on top of the image. The resulting display should look exactly the same as the image looked before you clicked on **Rasterize** in the last exercise.

Click on the **Select Color** button or press **<F3>**. **Dragon** pops up a dialog where you can type a color number and see the result. Type **150**. Then digitize several new roads using the same procedure as in the last exercise. Click the **Rasterize** to save the information you have created.

3. Choose **Geography⇒Polygons⇒Vector** again, enter the following parameters, and click **OK** to start the operation.

    - Image File: amhphoto.img
    - Apply Vector File: amherstRoads2.vec

The resulting display should be identical to what you saw before you chose **Rasterize**. **Dragon** added both the old and the new roads to the new vector file you created, **amherstRoads2.vec**.

4. Choose **Geography⇒Polygons⇒Measure**, enter the following parameters, and click **OK** to start the operation.

    - Image File: lamtaklong_110103_200103_b3.img

Notice that when you tab off the *Image File* field, **Dragon** fills in the *Units*, *X Scale* and *Y Scale* fields. This indicates that the selected image is georeferenced. This image is band 3 (red) of an Enhanced Thematic Mapper image, showing Lamtaklong Reservoir in Thailand.

Dragon displays the image you selected, in gray. It also displays a window holding a button panel and a text area, underneath the viewport. Using the same procedure as you used in **Geography⇒Polygons⇒Vector**, draw a polygon around the big reservoir in the upper left corner of the image. If you are using **Dragon Professional**, you may want to select a region that includes the reservoir first, so that you can see more detail.

When you click on the first point of the polygon, **Geography⇒Polygons⇒Measure** tags the point with a number. Pick points around the edge of the reservoir, then click **Close Polygon** to complete the process. Now click on **Measure** or press **<F8>**. **Dragon** displays the figure number, the perimeter, and the area in the text area to the right of the button panel.

Click the **Histogram** button or press **<F6>**. **Dragon** displays a new window with a graph that shows the number of pixels with each pixel value, for all the pixels inside your polygon. The histogram should show that most of the pixels have values of 32 or less. This makes sense, since water has very low reflectance in the red band. If your histogram shows many pixels with larger values, this suggests that you included some land pixels inside your polygon.

Create a polyline feature that starts in the upper left corner of the image, crosses the reservoir, and ends on the other side. Use **End Line** or **<F6>** to terminate the feature. Click **Measure** or press **<F8>** to measure the polyline. Notice that polylines do not have any area.

Now click **Profile** or press **<F7>**. **Dragon** displays a graph that shows how the pixel values change from the beginning of the polyline to the end. Notice how the values drop to near zero where the line crosses the reservoir, then go back to values mostly between 100 and 200. Do you see a spike where the line crosses the road to the east of the reservoir and then goes back into shadow?

Choose **Finished** to exit from **Geography⇒Polygons⇒Measure**. Unlike **Geography⇒Polygons⇒Vector** (and **Geography⇒Polygons⇒Cookie**), **Geography⇒Polygons⇒Measure** does not do any special processing before the operation ends.

5. Choose Choose **Geography⇒Raster⇒Buffer**, enter the following parameters, and click **OK** to start the operation.

   - Image File: amherstRoads.img
   - Distance Unit: Meters
   - X Scale: 10
   - Y Scale: 10
   - Target cell value: 150
   - Buffer distance: 100

This image is the one that you saved in the previous exercise using **Geography⇒Polygons⇒Vector**. It is not in fact georeferenced. However, the values we have provided for *X Scale* and *Y Scale* (basically, the pixel size) are approximately correct. We choose **150** for the target value because that was the value we used to draw the roads (and thus, the value that was saved in the rasterized layer).

Dragon computes the distance of every pixel from the closest pixel having a value of 150. Then it sets all pixels with a distance greater than the specified buffer distance (100 meters, or 10 pixels) to zero. The resulting image shows the target pixels themselves in white and the buffer area in red. All the red pixels are 100 meters or less from some part of some road.

6. Choose Choose **Geography⇒Raster⇒Slope**, enter the following parameters, and click **OK** to start the operation.

   - Image File: alfardtm.img
   - Viewport: 1

alfardtm.img is a 16-bit DEM from Alfaro, Spain. The area pictured has a deep gorge running between several high plateaus.

Dragon computes the slope at each pixel and displays the result as a gray scale image. Brighter pixels indicate steeper slopes.

Use **Display⇒Gray** to display the original image **alfardtm.img** in Viewport 0. In the DEM image, brighter pixels correspond to higher points on the ground. Compare the slope and the elevation images. Notice that the areas with the largest slope are at the edges, where the plateau plunges down into the gorge.

7. Choose Choose **Geography⇒Raster⇒Aspect**, enter the following parameters, and click **OK** to start the operation.

   - Image File: alfardtm.img
   - Color file: defaspct.clf
   - Viewport: 0

Dragon computes the aspect at each pixel and displays the result as a color image. Each color represents a different direction. There are eight possible directions,plus a null aspect for pixels that are on flat ground and have no slope.

Use **Display⇒Annotation⇒Legend** to place a legend on the image in Viewport 0. Notice that the dominant aspect in this image is Northeast. The stripes facing northeast are probably parallel ridges.

The floor of the gorge shows many different aspects mingled together. It is likely that this pattern is mostly noise. The slope image displayed previously suggested that the floor of the gorge is basically flat.

Chapter 7

UTILITY AND FILE OPERATIONS

7.1 Image Exploration and Management

Remote sensing images contain a great deal of information. In order to use that information most effectively, the analyst needs tools to examine the statistics and metadata for the image. The **Utility** and **Files** branches of **Dragon** supply several tools that allow you to explore your image data and manage your data processing.

The **Utility**⇒**Cursor** operation allows you to inspect data values in numerical form at any chosen location, for up to four different image bands. You can see how the values vary across bands, that is, the spectral signature of the selected pixel. This operation also provides an option that shows you values in a neighborhood around the chosen pixel position. You can use this to evaluate whether a particular region is homogeneous or not (and thus whether it would be a good location for defining a training area).

Utility also provides facilities for displaying one- and two-dimensional frequency graphs, often called *histograms* and *scatterplots*, respectively. A histogram shows the distribution of values in an image, that is, how many pixels have each possible value.

Two-dimensional frequency graphs or *scatterplots* are also a useful analytical tool. A scatterplot shows the relationship between image values at corresponding locations in two different images.

Other **Utility** operations perform a variety of useful tasks involving images and image files.

- List statistical, historical and identification information (*metadata*) from image file headers
- Modify historical and identifying information in an image file header
- Create and store a new color scheme

The **File** menu includes a variety of useful tasks involving files, both image files and other files, file paths, and program settings. Some **File** operations allow you to create new images and image files based on what is stored on disk or in memory. Other operations provide access to preferences and to the **Dragon** scripting capability. The **File** operations include:

- Save the current memory image in a new image file
- Create a spatial subset of an image file
- Script related functions including:
    - Running a script
    - Logging all executed commands to a text file which can later be used as a script
    - Turning on the script echoing function, which causes commands read from a script to be displayed in the History window as they are executed
    - Canceling a running script
- Operations for setting user preferences including:
    - The language to use in menus and messages
    - The paths to use as the defaults for reading data and writing data.
- Import image data in a variety of external formats into **Dragon** format.
- Exit from **Dragon**.

7.2   List Image File Header (LIS)

Choose **Utility**⇒**List** to display in tabular form information from image file headers. (See *Chapter 2* for descriptions of header information fields.) This includes identifying information, statistical information, and georeferencing information. For classified images, **List** also displays class names.

The information in many header fields can be changed using the **Header** operation, described in the next section.

List is one of the few operations that allow you to use a wildcarded file specification in order to indicate multiple files. For example, to list header information for all image files in the directory **D:\samples** whose names have the first three characters "**Ban**" you would type **D:\samples\Ban*** for the *Image File(s)* parameter.

If you use a wildcard to request the listing of multiple image files, **Dragon** will display header information only for the first 50 files that match the specification. To see the omitted files, make your wildcarded expression more specific.

The **List** command uses a different format for reporting image header information, depending on whether you enter a single image file name or a wildcarded image file specification. In the former case, the information will be formatted into a single page, with section delimiters for each category of information. In the latter case, the image information is displayed in a table, one line per file, with one column for each header data field.

Note that there is sometimes a slight delay between the time you click on the **OK** button and appearance of the file header report. This is normal.

7.3   Edit Image File Header (HEA)

Choose **Utility**⇒**Header** to modify the historical and identifying information stored in the header of an image file. Parameters are available for setting each user-controlled field in the header. Any fields for which you do not choose a new value remain unchanged.

The **File**⇒**Import** operation and the conversion programs supplied with **Dragon** (the *Tools Library* programs and **FCONVERT**) initialize these fields. If your image files are created in some other way you should use **Header** to set up valid headers for the new image files. **Header** calculates and stores image statistics, which **Dragon** uses to calculate default stretch parameters.

The first input parameter in **Header** is *Image File*. As soon as you specify a file name and move to another field, the other fields in the response panel will be filled in and will display the current values in that file. You may then modify the values as you like. The new values will be saved to the file when you press <**F1**>. Press <**Esc**> or <**F10**>, or click on **Cancel** to quit without modifying the selected image file header.

The input panel for **Header** has two tabs. The primary tab shows generic information for the image as a whole. The second tab allows you to enter or modify class names.

The capabilities available on this second tab are slightly different between **OpenDragon** and **Dragon Professional**. **OpenDragon** provides fields for seventeen labels, which can be associated with (up to) sixteen classes in a classified image (plus the value 0, which normally represents unclassified pixels). In **OpenDragon** the second tab will only be enabled for images of type **C**, that is, classified images.

In **Dragon Professional**, you can associate labels with any values, in any kind of image. For a classified image, these labels will normally be class names. For a layer file, the labels may be attributes associated with different data values. For a radiometric image, labels may identify breakpoints in a continuous scale. The second tab will be enabled as soon as you tab off the file name field in the first tab, regardless of the file type.

Also in **Dragon Professional**, the interface for data labels is organizd as a table. Click on **Add** to add a row to the table. Then enter the desired label and associated value. Values do not need to be consecutive. To remove a label, highlight its row and click on **Delete**. Click **Sort** to reorder the labels in terms of their values.

Header allows you to store a color file name in the header of any image file. If an image file has a color file name in its header, **Dragon** will generally use that color scheme for displaying that image, unless you specifically request a different color scheme. You cannot specify a path for the color file; **Dragon** assumes that the color file, if it is not one of the standard system color files, is in the same directory as the primary image file.

Header also allows you to store calibration function parameters in the image header. A *calibration function* is a function that maps the raw data values in the image file to external, more meaningful values. For example, a calibration function could be used to convert values from 0 to 255 into elevation measures. A calibration function has three components: the calibration offset, the calibration multiplier, and the calibration units. The calibrated values are calculated as follows:

```
calibratedValue = rawValue * multipler + offset
```

Calibrated values can be displayed using **Utility**⇒**Cursor**.

Most **OpenDragon** operations allow you to specify image files that are larger than the operating limits of the program. In this case, **Dragon** automatically uses a subset of the oversized image. **Utility**⇒**Header** is different; it will display an error if you specify the file name of an oversized image. Of course, this is not an issue in **Dragon Professional**.

7.4 Cursor Coordinates & Data (CUR)

Choose **Utility**⇒**Cursor** to find the exact line and pixel coordinates of specific points in an image and to display data values for up to four image bands at those coordinates. These functions are useful for defining subsets, for locating classification training areas, and for making correspondences between displayed images and other sources of geographic information such as maps or hardcopy images.

To use **Cursor**, you must first choose the image that you want to view while you are selecting points. This can be the memory image (=**M**) or the last displayed composite image (=**C**). You also specify from one to four image files whose data values you want to display. These files should all represent the same area and must have the same number of lines and pixels.

First, **Dragon** displays the image you request. The status bar (at the bottom of the image) shows the current image coordinates of the cursor. A secondary window appears containing a table which will be filled-in with values as points on the image are selected. This secondary window also contains buttons to end the processing, and to refresh the screen.

If you are using **Dragon Academic**, you can begin choosing points immediately. However, in **Dragon Professional** the image will initially be displayed as an *overview*. You must first choose a *region of interest*, as described in *Chapter 2*. The region of interest will then be displayed at a 100% (1-to-1) resolution. You may return to the overview (by clicking the right mouse button) and select a new region of interest as often as needed.

You can move the cursor using either the mouse or the cursor ('arrow') keys. The coordinate values in the status area will be updated as you move the cursor to from point to point.

When you press the left mouse button or the <**Ins**> key, **Dragon** displays an identifying number at the point you selected. The image files you specified are examined to obtain image data values for the current point. The secondary window will display all known information about the point specified for each of the images specified (but not for the background image). Information is displayed in recency order; that is, the most recently selected point is shown at the top of the table.

If an input image file has a calibration function specified (see **Utility**⇒**Header** for details), then the cursor operation will display the calibrated value, with the raw image data value in parentheses. Otherwise, only the raw image data value at the selected point will be displayed.

The numbers in the table column labeled **@ Geographic Coordinates** represent the geographic position of the selected point, as calculated based on the georeferencing information, if any, included in the image header. If no georeferencing information has been supplied, this values in this table column will be the same as the image coordinates. Georeferencing information may be added to an image using the **Header** operation, and is automatically stored when you register an image using **Geography⇒Register**. Note that, unlike the data values, the georeferenced coordinates are computed using information taken from the header of the *background* image. Thus, you cannot see georeferenced coordinates if you use **=C** (which is actually three images) as a background.

Cursor operates differently if you select a value other than zero (the default) for the *Window size* field. In this case, when you choose a point on the image, **Dragon** will display a table of image values for each data band, showing a small neighborhood around the pixel you have selected. The *Window size* you selected determines the width and height of this neighborhood. Each time you click on a new point in the image, the old neighborhood tables will disappear and a new table will be displayed.

To select a point more precisely, you may want to zoom (enlarge) the image on the screen. You can zoom using the viewport Zoom menu.

In a zoomed image, it is easy to see individual pixels. Thus you can locate the cursor very precisely. You use the left mouse button or **<Ins>** key to select a point and display its data values in a zoomed image, just as you would in a normal image. Note that the coordinates displayed refer to the image as a whole, not just the zoomed part currently visible.

When you want to return to the normal, non-zoomed image, choose the viewport **Zoom⇒Restore** operation.

To exit from **Cursor**, press **<F1>** or click on the **Finished** button. To clear previously selected points from the image, without exiting, press the **<F9 >** key or click on the **Refresh** button.

7.5   Histogram Display (HIS)

A histogram is a traditional way to show the distribution of values in an image; that is, how many pixels have each possible value. Figure 7.1 shows an example.

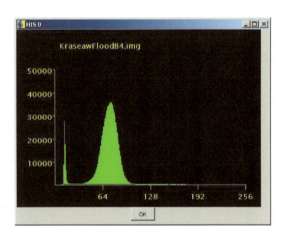

Figure 7.1: *Sample image histogram*

The X axis in this graph corresponds to different image values. This image uses one 8-bit byte for every pixel, so its values can range from 0 through 255. **Dragon** can also handle 16-bit images. A histogram of a 16-bit image will have an X axis ranging from 0 to either 16384 or 65535, depending on the maximum value in the image, divided up into *bins* that hold 256 values each.

The Y axis shows the pixel count, that is, the number of pixels/positions that each value. The maximum value on the Y axis will vary depending on the size of the image. An image that is 1024 pixels wide by 1024 pixels wide will have a maximum frequency of approximately one million.

Examining an image histogram can tell you a great deal about an image. The histogram will indicate whether the pixel values use the full available range of values (called the *dynamic range* or not. In Figure 7.1, almost all of the pixels have values below 128. This means that the image will have poor contrast and will be hard to interpret if it is displayed without histogram adjustment. You can use the histogram to decide on the optimum parameters for contrast adjustment. Then you can can apply those parameters by selecting the *User-defined Histogram adjustment* option. That option asks you to specify a *Lower Bound* and an *Upper Bound*. All pixels with values below the lower bound will be set to zero. All pixels with values above the upper bound will be set to 255. Pixels with values between the two bounds will have new values assigned so that the resulting display uses the full dynamic range. Normally, you should pick bounds that are close to the minimum and maximum values that have any significant frequency. For the image in Figure 7.1, good choices might be 16 for the lower bound and 100 for the upper bound.

A histogram can also indicate whether your image includes well-defined regions with different radiometric characteristics. The histogram in Figure 7.1 has two distinct peaks. The peak near the lower end of the dynamic range corresponds to a large reservoir. (This is the histogram of a near IR band, so water reflects very little radiation.) The other peak corresponds to the land area. You could use the histogram to decide on a value to use for thresholding the image (as discussed in Chapter 5). For this image, you should probably choose a value around 32, between the two peaks.

Histograms of classified images look somewhat different than radiometric images, as shown in Figure 7.2. Since **Dragon** supports only sixteen different classes in a single image, the X axis is labeled from 0 through 16. The frequency of each class shows up as a fairly thick bar.

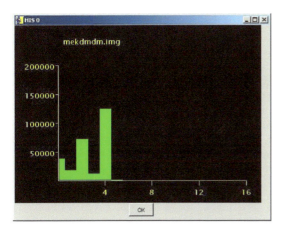

Figure 7.2: *Histogram of a classifed image*

Choose **Utility⇒Histogram** to display in graphic form the frequency distribution of data values in an image. **Histogram** adjusts for the fact that image values often cluster in the low end of the scale by magnifying the horizontal scale if necessary. The value of tic marks on the vertical axis varies depends on the maximum count represented in the histogram (which is used for scaling).

To display a histogram, you merely choose the image file you wish to use. You can choose to use one of the memory images, if they exist, but you cannot choose the composite (=**C**) image.

Histogram provides two additional controls using the fields *Apply Stretch* and *Mask image*.

Apply Stretch allows you to see the histogram of an image either unstretched, or with the most recently used stretch algorithm applied. This provides a means for you to see the effects on the histogram of any stretch you choose during a **Display** operation.

The *Mask image* field permits you to specify a mask image which limits the parts of the image used to calculate the histogram. **Dragon** calculates and displays a histogram based only on the pixels corresponding to non-zero locations in the mask image.

7.6 Scatterplot Display (SCA)

A *two-dimensional histogram* or *scatterplot* is a two-dimensional frequency graph showing the relationship between image values at corresponding locations (pixel positions) in two different images. Figure 7.3 shows an example.

The vertical axis in the scatterplot represents the different possible data values in the first image. The horizontal axis represents data values in the second image. Thus, each point in the graph space represents a particular pairing of values in the corresponding pixels of the two images. (These pairings are called *co-occurrence classes*.) A pixel position that had a high data value in both images would be 'located' in the graph space near the upper right corner. A pixel position that had a high value in the first image but a low value in the second would be 'located' in the upper left corner. For example, in Figure 7.3 there is a blue dot located around point (130,32), where the first coordinate is the X axis (**nakhon4.img**) and the second is the Y axis (**nakhon7.img**). This means that there is at least one pixel location where the value in **nakhon4.img** is 130 and the value in **nakhon7.img** is 32.

The frequency dimension of the scatterplot is represented by color. There are seven colors representing seven different frequency levels. White is the highest frequency; blue is the lowest. Unlike a histogram, a scatterplot provides only relative frequency information. Because the frequencies in a scatterplot can potentially be very large, **Dragon** scales the values before display. The scaling method preserves any co-occurrence class holding even one pixel. Thus areas of the scatterplot that are black indicate true zero frequencies; any co-occurrence class with even one pixel will show up in color.

Scatterplots are often used to evaluate whether two image bands are *correlated*. Bands that are highly correlated contain *redundant information*; the value of a pixel in one band can be predicted, to some extent, from the value in the other band. Typically, a scatterplot will show clusters, high-frequency groupings of pixels in specific regions of the graph space. If the shape of these clusters is circular, or elliptical with the main axis of the ellipse horizontal or vertical, this indicates that the two image bands being graphed carry largely independent information and that both bands should probably be used in a classification. On the other hand, if the scatterplot shows mostly clusters with axes pointing diagonally toward the corners of the graph space, the two bands are highly correlated and carry largely redundant information. In this case, it may not be advisable to use both bands in a classification.

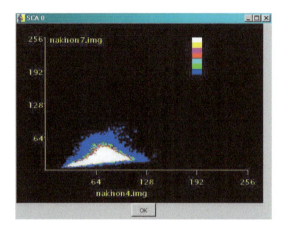

Figure 7.3: *Sample scatterplot display*

The scatterplot in Figure 7.3 shows limited correlation between bands 4 and 7 of this image.

In addition to scatterplots based on full images, it is often useful to view scatterplots of training areas chosen for classification. A good set of training areas should produce nicely separated clusters in a scatterplot. If training areas overlap or merge together in spectral space, the classification process will not be able to unambiguously separate those two classes. **Dragon** does not offer a "Signature Scatterplot" operation. However, signature scatterplots can be created by using masking to exclude all pixels from the input except those that were chosen for training. The exercises at the end of this chapter provide a step-by-step example of this process.

Choose **Utility**⇒**Scatterplot** to display in graphic form the joint frequency distribution of data values in two images or image bands.

Since image data values frequently cluster in the lower ends of the reflectance scale. **Dragon** automatically enlarges the scatterplot scale two or four times, showing only the lower ends of the image data scales when there are no points located in the higher ends of the scales. Both axes are adjusted by the same amount, so that the band with the larger dynamic range will determine the magnification factor.

To display a scatterplot, you merely choose the two image files you wish to use. Usually these will be two bands from the same scene, though this is not required.

7.7 Color Scheme Creation (COL)

Choose **Utility**⇒**Color** to create a new color scheme for assigning colors to image data values, and to save that scheme. **Color** can also be used simply to view an existing color scheme.

A **Dragon** color scheme assigns one color to each possible eight bit data value (0 through 255). In creating a color scheme, **Dragon** starts with a base color scheme. Any color-to-value mappings that you do not explicitly change will remain as set in the base color scheme.

In addition to specifying a base color scheme, you can also enter the name of a color file for saving your new color scheme. If you do not enter a second color file name, you can view the contents of an input color file, but not change it.

Finally, you may select a classified image that corresponds to the color scheme you are creating. If you do specify a classified image file, **Dragon** will locate that file, read its header, and display a table similar to the legend which can make it easier for you to assign colors to classes.

When you have finished entering parameters for **Color** and click on **OK** or press <**F1**>, **Dragon** displays a color chooser dialog. If you specified a classified image file to use as a reference, this dialog has three panels as shown in Figure 7.5 otherwise, it has two, as illustrated in Figure 7.4.

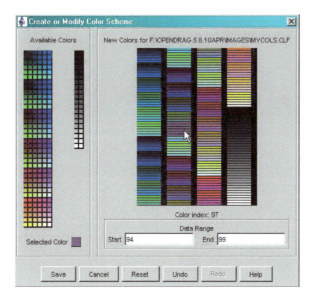

Figure 7.4: *Color chooser dialog with no classified image*

The leftmost panel shows two grids, one with colors and one with shades of gray. This panel displays all the available colors. To choose a color or gray shade, click on the corresponding square with the mouse. The box labeled 'Selected Color' will change to the color you selected.

The middle panel provides several ways that you can assign the selected color to one of more data values. This panel shows an array of horizontal colored bars. If you specified a classified file, there will be seventeen bars; if you did not, there will be 256, arranged in four columns of 64. Each bar shows the current color of the corresponding data value.

As you move the mouse over these bars, you'll note that the field labeled 'Color index' changes to show you the data value associated with each bar. This field changes to a set of dashes when the cursor is outside the color display area.

The panel also includes two text fields for entering a data range start and end value.

To assign the selected color to a single data value, do either of the following:

- Move the mouse until the cursor is on top of the data value you want (as indicated by the *Color index* field), and click with the left mouse button.
- Use the mouse to select the *Data Range: Start* field and type in the value you want. Tab to the *Data Range: End* field and type the same value. Then press **<Tab>** once more.

Either of these action sequences will cause the corresponding color bar to change to the selected color.

You can change a range of data values to the selected color, as well. Either type in the start and end data range values in the text entry fields, or else place on the starting value in the color bar array and press the left mouse button, drag to the ending value, and release the left mouse button. In either case, the entire range of values will change to the selected color.

The Color Chooser dialog allows you to undo the last change by clicking the **Undo** button. If you want to reapply the last change, click on the **Redo** button. To discard all changes since you began a **Color** session, click on **Reset**. (Note that a Reset operation cannot be undone.)

If you specify a classified image as an input parameter to **Color**, the Color Chooser contains a third panel, on the right. This panel lists the class names associated with each class data value, plus the color associated with that value in the base color scheme. As you select values in the middle panel, the right panel also changes, so you can easily see what color will be associated with each class if you use the modified color scheme to display the classified image.

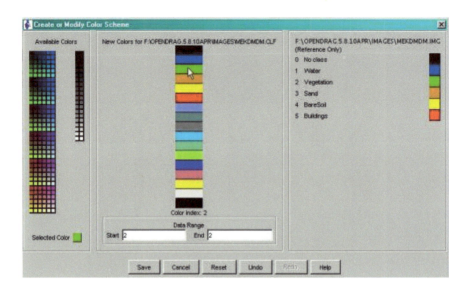

Figure 7.5: *Color chooser dialog with classified image*

When you are satisfied with your new color assignment, click on the **Save** button to write that scheme to the file you specified. Note that if this was a previously existing file, it will be overwritten with the new values. Click on **Cancel** to exit without making any of your changes permanent. You can use the **Cancel** option if you simply want to view a color file, without making any changes.

Once you have defined and saved a new color scheme, you can use it in Display or Enhancement operations by choosing the value *Retrieve* for the *Color Option* parameter and then typing the name of your new color scheme file for the *Color File* parameter. If you want the color scheme to be used automatically whenever your classified image is displayed, use **Header** to associate the color scheme with the image file by storing the color file name in the header.

7.8 Save Current Image (SAV)

Choose **File**⇒**Save** to store permanently on disk the image currently in memory image $=$**M**. Normally, this image will be the result of the most recent enhancement, classification, or display operation. If there is no image currently in $=$**M**, you cannot choose **Save**.

One common use of **Save** is to preserve intermediate results that you intend to use as input to later operations. For example, to calculate a vegetation index, with the formula

    (Band 7 - Band 5) / (Band 7 + Band 5)

you could calculate the band sum first and save the result, using **Save**. Then you can calculate the difference, and finally use your saved image file as the denominator in forming the ratio.

Note

You could perform these calculations without creating an intermediate file, using the **Enhance**⇒**Vegetation** operation. The computation is discussed here only to explain the general process of using **Save** for intermediate results.

Most fields in the image header will be set automatically during the **Save** process. You can enter a descriptive comment to indicate where the image came from and what processing it has received. If desired, you can modify the other header information fields later using **Utility**⇒**Header**.

You can use **Save** to save the immediate results of a classification. Initially the class names will be the names you gave to the signatures in a supervised classification. For unsupervised classification results, the class names will be strings of the form *Class 1*, *Class 2* and so on. To change the class names, you must save the classified image first and then use **Utility⇒Header**.

Save calculates image statistics and stores them in the image header.

7.9   Subset Creation (SUB)

Choose **File⇒Subset** to extract part of an existing image file and load it into image memory. **Subset** can be used to create a new image file that holds only part of an existing file. In **OpenDragon**, **Subset** provides one way to view any portion of an image file that has too many lines or pixels per line to be displayed at full resolution all at once.

To use **Subset**, you must know the starting line and pixel for the subset you want to extract, as well as the number of lines and pixels to be in the subset. You can use the Utility operations **Utility⇒Cursor** and **Utility⇒List** to help you to find this information.

Dragon reads the header of the image file that you want to subset and checks the parameters against the size of the file. It notifies you of any inconsistencies and requires that you make them consistent. Note that **Dragon** will not allow you to create subsets smaller than 256 by 256.

Subset copies the selected section of the image file you request into the main memory image (=M), and then displays the resulting subset. You can make the subset into a permanent image file with **File⇒Save**.

The **File⇒Import** operation discussed below also provides a subsetting capability on the **Custom** tab. The main difference is that **Subset** loads the resulting image into image memory where it can be further manipulated. To create a new image file, you must use **File⇒Save**. In contrast, **File⇒Import** creates an output image file, but does not load the subsetted data into image memory.

7.10   Script Menu

The **File⇒Script** menu includes operations for invoking and controlling scripts. A script is a text file that contains commands in the **Dragon** command language. See the *Scripts and Immediate Commands* chapter of the *User Manual* for a detailed discussion of scripts. Basically, a script provides a way of automating a sequence of processing steps.

The **Script** operations differ from most of **Dragon** operations in that they do not generally display response panels. Instead, they affect the overall state of the system.

Run Command

File⇒Run starts a script running. When you select **Run**, **Dragon** displays a File Dialog which allows you to select the script file to run. When you click on **OK** in this dialog, the script begins to run. If you click on **Cancel** instead, no script is run.

Log Command

Choose **Log** to turn on or off logging (that is, copying and saving to a file) of commands being executed by **Dragon**. Logging is useful because it provides a history of your work during a session. A log file also provides a good starting point for developing a script file. In fact, many log files can be run as scripts without any modification.

When you select the **Log** menu item, **Dragon** displays a file dialog which allows you to navigate to the desired output directory and enter or select a file name. If you select an existing file, new logging output will be appended to that file. When you click on the **OK** button after selecting or entering a filename, **Dragon** records its state as *logging*. After this point, if you display the **Script** submenu, you will see that the **Log** item is checked.

To turn off logging, simply select the **Log** item a second time. The check mark will disappear, indicating that **Dragon** is no longer logging command input.

Echo Command

Choose **Echo** to turn on or off the script echoing functionality. When echoing is turned on, all commands read from a script file are displayed in the History window of the **Dragon** menu client as they are executed. This makes it easy to keep track of the status of a script. Commands read from a script are enclosed in parentheses in the History display; commands derived from interactive use of **Dragon** are not.

The **Echo** menu item requires no input. If you select it, **Dragon** will display it as checked, meaning echoing is enabled. If you select it a second time, the check mark will be removed, indicating that echoing is disabled.

Note that the menu item has no effect on a running script. In fact, when a script is running, the menu item is disabled.

Cancel Command

Choose **Cancel** to cancel a running script. It has the same effect as clicking on the **Cancel** button on the Script Control Panel. (See the *Scripts and Immediate Commands* chapter for a details about the Script Control Panel.)

Cancel is enabled only when **Dragon** is in the process of executing a script. When **Cancel** is chosen, **Dragon** will complete its processing of the current command before ending the script.

7.11   Preferences Menu

The **Preferences** menu includes items that allow you to customize your **Dragon** installation.

Language

When you choose **File**⇒**Preferences**⇒**Language**, **Dragon** displays a dialog with multiple radio buttons. Each radio button corresponds to a different language; in general the name of the language will be specified using the appropriate word and characters for that language.

Some radio buttons may be disabled; these indicate languages for which no translation files are available in the current installation. The currently selected language will be indicated by the fact that its radio button is selected.

To change the language **Dragon** uses, select a different radio button and then click **OK**. Your selection will take effect the next time that you run **Dragon**. So, if you want the new language immediately, you should exit **Dragon** and then restart.

The currently selected language is recorded in the configuration file, **dragon.cfg**. You can change the language without running **Dragon** by editing this file and changing the setting of the $COUNTRY=$ variable. Currently supported country settings include *EN* for English, *BA* for Indonesian (Bahasa Indonesia), *CZ* for Czech, *FR* for French, *PT_BR* for Brazilian Portuguese, *RU* for Russian, and *TH* for Thai.

If you exit from the language selection dialog by clicking on the **Cancel** button, the language will not be changed.

Paths

When you choose **File**⇒**Preferences**⇒**Paths**, **Dragon** displays a response panel showing the current values for the following items:

- *Default image path.* This is where **Dragon** looks, by default, for image files, color files, polygon files, vector files and so on. (In previous versions of **Dragon**, this was known as **-DPATH**.)

- *Default file creation path.* This is where **Dragon** will write its files by default: image files created with **File⇒Save**, vector files created with **Geography⇒Vector**, signature files created by training or editing, etc. (In previous versions of **Dragon**, this was known as **-WPATH**.)

These settings will take effect as soon as you click on the **OK** button.

Often it is desirable to have the *Default image data path* and *File creation path* be set to the same directory. This makes it easy to create files in one operation and read them in another. However, there are circumstances when you might want these directories to be distinct. One common situation would be if your primary image data files were located on a network drive or on a CDROM.

Of course, **Dragon**'s file dialogs allow you to browse and select files for input or output from any accessible drive and directory. Specifying the defaults intelligently, however, makes it easier to enter file names directly into the response panel fields. For instance, if you enter an input image file without a path, **Dragon** will automatically prefix the file name by the default image path and look in that directory in order to verify that the file exists.

7.12   Import (IMP)

Choose **File⇒Import** to convert image data from external formats to **Dragon** image format.

To use **Import**, specify the necessary parameters as described below and then click on **OK** or press **<F1>**. **Import** does not display any images or messages, unless an error occurs. After you start the operation, the user interface will indicate that **Dragon** is "Working..." for a few minutes. Then the response panel will disappear. This means that **Import** has imported the files that you requested, creating new Dragon image files with the names that you specified.

When you choose **Import**, you will see a response panel with two tabs, labeled **Automatic** and **Custom**.

Automatic Import

When you use the Automatic tab, you simply specify the name of the input file that you want to import, and a prefix for the output file name. **Dragon** figures out whether the input file is a format that it can convert. If the input file is a known format, **Dragon** will create one or more **Dragon** format image files from the data in the input file.

Formats that can be imported using the **Automatic** tab include:

- **Dragon** image files (but for these files, import is not necessary)
- Single band TIFF or GeoTIFF files
- Multiple band TIFF or GeoTIFF files
- ESRI BIL files
- USGS 7.5 minute DEM (Digital Elevation Model) files

There may be additional formats.

In addition to the input file name and output file prefix, the automatic import panel offers two other parameters. The *Band to extract* allows you to convert only one, selected band from a multi-band TIFF or BIL input file. If you leave this parameter set to its default value of zero, **Dragon** will create a separate image file for each band. These files will be named *prefix-#*.img, where *prefix* is the specified output prefix and the number sign is replaced by the band number, starting at 1.

The second parameter is *Reduce size*. If you check this box, **Dragon** will automatically determine the necessary downscaling factor, and will subsample the input data so that the output file can be displayed in a window of 1024 x 1024 or less. For example, if the input image was 2048 lines x 3072 pixels, **Dragon** would copy every third line and pixel to the output file, resulting in an image that was 682 x 1024.

If you are using **OpenDragon**, *Reduce size* is useful if the input image has larger dimensions than **Open-Dragon** can display. (**Dragon Professional**, of course, can correctly display images of essentially unlimited size so *Reduce size* is not of much use unless you are preparing images for students who will be using **Open-Dragon**).

Custom Import

The **Automatic** tab of the **Import** operation requires only a small amount of information, but also gives you only a limited amount of control over the output. The **Custom** tab provides a far wider range of options and parameters.

You should use the **Custom** tab when:

1. You want to import an image that is not one of the types automatically recognized by **Dragon**;

2. You want complete control over subsetting or subsampling during data import.

In order to import a foreign image from a format that **Dragon** does recognize, you need to know the details of the format and of the particular file. In particular, you need to know the dimensions of the image (number of lines and pixels per line), number of bands, and the size of any non-image data such as a header that precedes the image data.

The **Custom** tab includes four groups of parameters.

- Input image file name and output file prefix: These identify the image you want to import and determine the resulting name, in a manner similar to the **Automatic** branch of **Import**.
- Image type and operation: The possible image types are either **Dragon** (which includes all the types recognized automatically) or **Foreign**. You should use **Dragon** for any of the image types that **Dragon** can import automatically (see above). If you specify that the file to be imported is **Foreign**, you will need to provide additional information on its structure using the Input Image Description parameters. The four possible operations are **Copy**, **Subset**, **Reduce**, and **Extract band**. Depending on which of these you select, you will be required to enter values for some of the Output Parameters.
- Input image description: These parameters tell **Dragon** the format of your foreign file. They include *Lines*, *Pixels*, *Header size*, *Number of bands*, and *Data size*. The fields for these parameters will be enabled only if you select **Foreign** as the *Image type*.
- Output parameters: These parameters are enabled if you choose an operation other than **Copy**. They allow you specify the details such as the size of a subset, the reduction factor, or the band to extract. Different output parameters will be enabled depending on the operation that you select.

For **Dragon** format images, using **Custom** import is nearly as simple as using the **Automatic** tab. For foreign image files, the main difficulty is figuring out what values to supply for the Input Image Description parameters. The **Custom** tab can successfully convert many image formats provided that they have the following characteristics:

- All lines have the same number of pixels and all bands have the same number of lines. That is, the image data must be 'rectangular' for all bands.
- Each pixel is represented by either 8 or 16 bits of binary data, interpreted as unsigned bytes or short integers. Data files that use ASCII text to represent pixel values cannot be converted. Floating point data is also outside the capabilities of the current system.
- The image data is not compressed in any way. For instance, run-length-encoded files cannot be converted.
- If the image is multi-band, pixels for different bands are arranged in either pixel interleaved, line interleaved (also called 'band interleaved'), or band sequential organization. (See the Figures 7.6 through 7.8 for an explanation of these terms.)
- If an image has any kind of header preceding the data, this header is of a predictable and constant size.
- If an image has any kind of control data at the beginning or end of lines, this data is of a predictable and constant size.

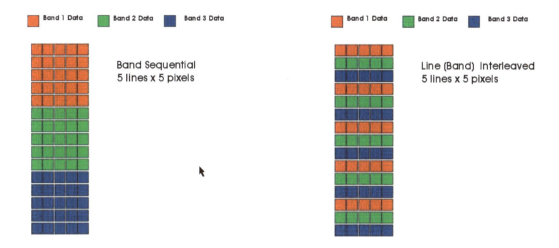

Figure 7.6: *Band sequential layout*   Figure 7.7: *Band interleaved layout*

To copy the data, subset the data or subsample the data from a foreign file that contains a single band, you must enter the appropriate input image description information. Then you can proceed as if you were working with a **Dragon** format image file.

To work with a foreign file that contains multiple bands, you must supply 'fake' input image description data that treats the input as a single rectangular block of data. Then you select the specific band that you want by specifying appropriate subset or reduction parameter values. Note that you cannot use the **Extract band** operation with a foreign file type.

If your foreign image has a band sequential format, specify the actual number of lines times the number of bands as the line count in the input image description. Specify the actual number of pixels per line. For the band sequential example in Figure 7.6, you would specify 15 lines and 5 pixels per line. To extract a particular band, choose the **Subset** operation. Specify the first line in the band desired as the starting line of the subset: 0 to extract the first band, 5 to extract the second band, 10 to extract the third band, in the case of the Figure 7.6 example. Specify 0 as the starting pixel value, the real number of lines per band (e.g. 5) as the number of lines for the subset, and the real number of pixels per line (e.g. 5) as the number of pixels in the subset.

If your foreign image has a band interleaved format, in the input image description specify the correct number of lines, but a number of pixels per line that is equal to the number of pixels per line times the number of bands. For the example in Figure 7.7, you would specify 5 lines and 15 pixels per line. To extract a particular band, choose the **Subset** operation. Specify line 0 as the starting line of the subset in all cases. Specify the first pixel position for the band that you want to extract as the starting pixel of the subset. In the example above, you would specify 0 to extract the first band, 5 to extract the second band, and 10 to extract the third band, Specify

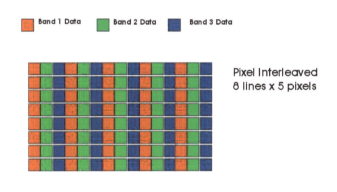

Figure 7.8: *Pixel interleaved layout*

the real number of lines per band (e.g. 5) as the number of lines for the subset, and the real number of pixels per line (e.g. 5) as the number of pixels in the subset.

115

If your foreign image has a pixel interleaved format, in the input image description specify the correct number of lines, but a number of pixels per line that is equal to the number of pixels per line times the number of bands. For the example in Figure 7.8, you would specify 8 lines and 15 pixels per line. To extract a single band, choose the **Reduce** operation. The reduction factor should be equal to the number of bands (3 in the example above).

This procedure will extract the first band from a pixel interleaved format file. To extract the second band, you must tell the software that the input file has a header that is one byte in size. (This assumes that the file contains 8 bit data.) To extract the third band, indicate that the file has a two byte header.

If you are not sure what parameters to provide in converting a foreign image, drawing a picture of the data organization will sometimes help. Also, you can try different alternatives and check the results. Incorrect parameters will usually result in output images that are all black, are scrambled so that no image features are visible, or that look squished in the horizontal or vertical direction.

7.13   Exercises

This section offers a set of hands-on exercises with **Dragon** utility and file operations, in order to let you practice using these functions and see their results.

1. Start your **Dragon** system. Choose **Utility**⇒**Scatterplot**. Enter the following parameters:

   - Image File 1: losan1.img
   - Image File 2: losan2.img
   - Skip factor: 1

   Click the **OK** button to start the operation. **Dragon** displays a scatterplot showing the joint frequency distribution of these two image bands. Notice that most of the pixels form a narrow clump that slopes diagonally from the lower left to the upper right of the graph. This indicates that the first and second bands (green and red) of this image are strongly correlated. Leave this scatterplot visible while you go on to the next exercise.

2. If you have not done Exercise 10 in *Chapter 4* (principal components analysis exercise), you should do so before proceeding.

   Choose **Utility**⇒**Scatterplot** again. Enter the following parameters:

   - Image File 1: losan-C1.img
   - Image File 2: losan-C2.img
   - Skip factor: 1

   The resulting scatterplot is very different from the one displayed in step 1. The data points form a cloud that covers a much wider range in both dimensions. The main direction of the cloud is parallel to the X axis, indicating that the data values are largely independent. Indeed, one of the major reasons for performing a principal components transformation is to produce bands that are mostly uncorrelated. (See *Chapter 4* for further discussion.)

   Remember that we performed a multiplicative scaling in the principal components operation. This results in some *saturated* pixels, that is, pixels whose values are truncated at 255. The lines at the top and right sides of the graph reflect this. The fact that the lower right corner of the graph is empty is also an artifact of the scaling. You may wish to experiment with other scaling options in the principal components exercise to see how they affect the scatterplot.

3. Choose **Display**⇒**3 Band**. Enter the following parameters:

   - Image File for Blue: mek14-1.img
   - Image File for Green: mek14-2.img
   - Image File for Red: mek14-3.img
   - Viewport: 0

   Click the **OK** button to start the operation. **Dragon** displays the three band composite of the Mekong River which we used in *Chapter 5*. Now choose **Display**⇒**Overlay**, specify the following parameters, and click **OK**:

Copyright ©Global Software Institute

- Image File: mekongTrainMask.img
- Histogram adjustment: None
- Viewport: 0

The image **mekongTrainMask.img** was created using **Geography**⇒**Vector**. It is intended to identify training areas for five classes: water (dark blue), vegetation (light green), buildings (cyan), bare ground (brown) and sandbar (yellow). We can use this mask image with the **Classify**⇒**Training**⇒**Automatic** operation. This operation makes it possible to select training areas that are irregular polygons rather than just circles. See Chapter 5 for details.

4. Before we extract signatures, however, we would like to use the **Utility**⇒**Scatterplot** operation to evaluate whether the training areas are independent. To do this, we need to create masked versions of the input bands. Choose **Enhance**⇒**Mask**, enter the following parameters, and click **OK**:

- Image File 1: mek14-1.img
- Image File 2: mekongTrainMask.img
- Histogram adjustment: None
- Viewport: 0

Using **File**⇒**Save**, save the resulting image as **mektraining1.img**. Now choose **Enhance**⇒**Mask** again, enter the following parameters, and click **OK**:

- Image File 1: mek14-3.img
- Image File 2: mekongTrainMask.img
- Histogram adjustment: None
- Viewport: 0

Using **File**⇒**Save**, save the resulting image as **mektraining3.img**.

5. Now choose **Utility**⇒**Scatterplot** enter the following parameters, and click **OK**:

- Image File 1: mektraining1.img
- Image File 2: mektraining3.img
- Skip factor: 1

Dragon displays a scatterplot that shows only the pixels in the training areas, for bands 1 and 3 of this SPOT subscene. Notice that the graph shows five distinct clusters, once for each class, with very little overlap. This indicates that our training areas are in fact fairly unambiguous. At the same time you will note that there a few points in the upper right part of the graph that do not seem to belong to any cluster. These are possibly errors.

6. Choose **Utility**⇒**Cursor**, enter the following parameters, and click **OK**:

- Image File to display: mekongTrainMask.img
- Band 1 Image File for data: mek14-1.img
- Band 2 Image File for data: mek14-2.img
- Band 3 Image File for data: mek14-3.img
- Window size: 7

Dragon displays the mask image. Select a point in one of the blue polygons (water areas). (If you are using **Dragon Professional** you will need to select a region first.) In the text area below the image, **Dragon** displays the neighborhood data values from the three Mekong image bands. The value of the selected point is in bold, in the center of the tables. In each band, the values should be fairly consistent. Make note of the central pixel value in each band.

Now select a point in one of the green polygons (vegetation areas). Once again, **Dragon** displays the neighborhood data values. Notice that they are quite different. Band 1 is fairly similar to the water class, but band 2 is quite a bit lower and band 3 much higher than the values for water. Continue to explore the signatures for different classes. The **Utility**⇒**Cursor** operation allows you to examine image data values in great detail.

7. Choose **Utility**⇒**List**, enter the following parameters, and click **OK**

- Image File: lamtak*

Dragon displays the metadata for all files in your default data directory that begin with "lamtak". You should see at least eight files, four wet season image (**lamtaklong_110103*.img**) and four dry season (**lamtaklong_250205*.img**). Notice that the size of these images is 1777 lines by 1876 pixels, too large to be displayed completely by **OpenDragon**.

Choose the **Log** item under **File⇒Script**. When the system prompts you to enter a log file name, enter **subset.spt**. Now whatever operations you execute will be captured in command-line format to that file.

Choose **File⇒Subset**, enter the following parameters, and click **OK**

- Image File: lamtaklong_110103_200103_b1.img
- Initial Line: 0
- Initial Pixel: 0
- Number of Lines: 1024
- Number of Pixels: 1024

Dragon extracts the specified subset from the band 1 image and displays it in gray in Viewport 0.

Choose **File⇒Save** and save the result as **LamtakWetSub1.img**.

Choose the **Log** item under **File⇒Script** again. This turns logging off. Now open the file **subset.spt** in Notepad or some other text editor. (The file will be located in whatever directory you specified as the default output directory when you installed **Dragon**. Notice that the file contains the commands to create the subset and then save it.

In your editor, copy the contents of the two commands and then paste them twice. Now, in the first set of commands, change the filenames: change "b1" to "b2" in the **SUB** and change "1" to "2" in the **SAV** command. Repeat this for the two copies you pasted, changing "1" to "3" and "1" to "4" respectively. Save the script file and exit.

To run the script you just created, select **File⇒Run**. In the file chooser, select your script **subset.spt**. Click on **OK** to start the script running. Notice that **Dragon** displays three images in quick succession, as it creates subsets for bands 2, 3 and 4 of the input image.

Choose **Display⇒3 Band**, enter the following parameters, and click **OK**

- Image File for Blue: LamtakWetSub2.img
- Image File for Green: LamtakWetSub3.img
- Image File for Red: LamtakWetSub4.img

Dragon displays a three band composite using the subsetted images that you created with your script.

Choose **Utility⇒List** again, enter the following parameters, and click **OK**

- Image File: lamtak*

You should still see the original 1777 by 1876 source images. You should also see the four subset files that you created: **LamtakWetSub1.img** through **LamtakWetSub4.img**. Note that the dimensions of these files are all 1024 lines by 1024 pixels (as you specified).

Chapter 8

BIBLIOGRAPHY

8.1 Books

There are many books on remote sensing and geographic information systems which can provide background for **Dragon** users. This list focuses on books with which we have personal experience.

DeMers, M.N., *Fundamentals of Geographic Information Systems,* New York: John Wiley and Sons, 2002. Highly readable introduction to concepts and applications in geographic information systems, suitable for undergraduates.

Gibson, P.J. and Power, C.H., *Introductory Remote Sensing: Digital Image Processing and Applications.,* London: Routledge, 2000. Textbook focused on examples, with datasets and sample software, suitable for high school or undergraduate readers. This book includes a limited edition of Dragon/ips (r), but unfortunately it is an older version suitable only for Windows98 or MS-DOS computers.

Gonzalez, R.C. and Wintz, P., *Digital Image Processing,* Reading, MA: Addison-Wesley, 1987. Broad treatment of image processing techniques supported by both clear examples and mathematical background, suitable for graduate students and professionals.

Jain, Anil K., *Fundamentals of Digital Image Processing,* Englewood Cliffs, NJ: Prentice-Hall, 1989. A broad, mathematically-oriented treatment of image processing algorithms, suitable for graduate students and professionals.

Jensen, J.R., *Introduction to Digital Image Processing: A Remote Sensing Perspective,* Englewood Cliffs, NJ: Prentice-Hall, 1986. Textbook aimed at undergraduates with effective examples and figures.

Lillesand, T.M. and Kiefer, R.W., *Remote Sensing and Image Interpretation,* New York: John Wiley & Sons, 1994. Comprehensive and popular textbook on remote sensing which covers visual as well as digital interpretation techniques.

Schowengerdt, R.A., *Techniques for Image Processing and Classification in Remote Sensing,* New York: Academic Press, 1983. Somewhat old but concise and well-written treatment of fundamental digital image processing algorithms.

Tomlin, C.D., *Geographic Information Systems and Cartographic Modeling,* Englewood Cliffs, NJ: Prentice-Hall, 1990. An early and highly influential treatment of raster geographic information systems analysis.

8.2 Journals

You will find a wide range of journals, both on-line and off-line, which publish articles related to remote sensing and geographic information systems. Some titles are listed below, along with the names of the publishing organizations if relevant.

Many of these journals have some or all of their content available on-line.

These journals vary in their level of technical detail and target audience. Some are focused on publishing the latest research. Others are more application-oriented. Some are aimed primarily at the scientific community. Others are designed more for practitioners or even the general public. In the list below, we have tried to code the journals with which we are familiar, as follows:

- **R** - Research-oriented
- **A** - Application-oriented
- **S** - Scientific audience
- **P** - Popular or practitioner audience

- Asian Journal of Geoinformatics (Asian Association on Remote Sensing) (**RS**)
- Cartography & Geographic Information Society
- Asian Journal of Geoinformatics (Asian Association on Remote Sensing) (**RS**)
- Cartography & Geographic Information Society
- Canadian Journal of Remote Sensing (Canadian Remote Sensing Society)
- Computers & Geosciences
- EARSeL Newsletters
- Earth Observation Quarterly (EOQ) (**AP**)
- Geographic Information Sciences
- Geo Info Systems Magazine (**AP**)
- GeoInformatica
- GeoWeb Online Articles
- GIS World (**AP**)
- International Journal of Geographical Information Science
- IEEE Transactions on Geoscience and Remote Sensing (IEEE Geoscience and Remote Sensing Society) (**RS**)
- International Journal of Remote Sensing (**RS**)
- Journal of Geographic Information & Decision Analysis
- Journal of Geographical Systems
- Journal of Photogrammetry and Remote Sensing (ISPRS) (**RS**)
- Journal of Spatial Science
- Photogrammetric Engineering & Remote Sensing (PE&RS) (American Society for Photogrammetry and Remote Sensing) (**R,AS**)
- Photogrammetric Journal of Finland
- Remote Sensing of Environment
- Terra Forum
- URISA Journal (**R,AP**)

Index

Histogram, 92
History, 21
left mouse, 23, 25, 55, 56, 80, 83, 89, 92, 105, 106, 109
Measure, 91, 92
mouse, 24
Move Down, 98
Move Up, 98
New Class, 22, 55, 56
OK, 23, 25–27
Profile, 92
Rasterize, 80, 89, 90
Redo, 109
Refresh, 106
Region, 89–93
Reset, 109
right mouse, 24, 55, 80, 90, 92, 93, 105
Save, 109
Select Color, 89, 90
Sort, 104
Status, 21
Undo, 109
Yes, 68

CAL, *see* Calculate Registration Coefficients
CAL results report, 85
Calculate Aspect from Elevations, 80, **94**
Calculate Registration Coefficients, 20, 81, 82, 84, **84**
Cancel button, 27
CANCEL command, **112**
cancel current operation, 28
cell value, **3**
change class values, 69
channel file, **68**
channels, **4**
class
    change values, 69
    define names, 55
    map, 53
    names, 17, 36, 58, 60
    recode values, **69**
    view names, 58
class map
    image, 57, **57**
classes
    recode, 13
classification
    boxcar, 53, 59, **62**
    function summary, 53
    maximum likelihood, 53, **64**
    minimum distance to mean, 53, **66**
    multispectral, 54, 63
    number of bands, 62, 65, 66
    operations, **53**
    parallelepiped, 53
    save results, 53, 110

    supervised, 13, 53, 57, 61
    unsupervised, 13, 53, 67
Classification Error Report, **70**
classified image, **16**, 31
    color scheme, 69, 108
    display legend, **36**
    enhancement, 40
    file, **16**, 17, **54**
    histogram, 106
Close Polygon, 89
Close Polygon button, 80, 89–93
cloud threshold, 68
CLU, *see* Cluster Classification
Cluster Classification, **67**
clustering, 13, 53, 57
    number of bands, 67
    number of classes, 67
co-occurrence classes, 107
coded image, **6**, 30
coefficient file, **20**, **81**
COL, *see* Create/Edit Color Scheme
color
    composite, 13, 32, 34
    control of, 13
    pseudo, 13, **32**, **34**
    vector drawing, 88
color file, **19**, 54, 110
    name in header, 17
    names, 19
color scheme, 19, 31, **54**
    AGC, **68**
    Ambroziak, **68**
    base, 108
    classified image, 69, 108
    create, 19
    definition, **19**
    gray, 34
    image, 108
    modify, 19
color vs. gray-scale display, 32
COM, *see* Combine Signatures
COMB, *see* Combine Layers
Combine Layers, 80, **96**
Combine Signatures, **60**
command
    -DPATH, 21, 112
    -WPATH, 21, 113
    1BA, 16, 32, 34, **34**, 68
    3BA, **34**
    ADD, 35, **35**
    AGC, **68**
    ANN, **35**
    APP, 54, **56**, 57
    ASP, 80, **94**
    Automatic Signature Generation, **57**

122

Special discount for readers of this book!
Purchase your own personal copy of
Dragon Professional for only $395.
Please supply the following discount code
when ordering:

GSI3.141

To order visit *http://www.dragon-ips.com*
Offer valid at least until December 2012.

www.ingramcontent.com/pod-product-compliance
Lightning Source LLC
Chambersburg PA
CBHW041418050326
40689CB00002B/559